CARS
Champions of Style

pil

Publications International, Ltd.

Louis Weber, CEO
Publications International, Ltd.
8140 Lehigh Avenue
Morton Grove, IL 60053

Permission is never granted for commercial purposes.

ISBN: 978-1-64030-649-3

Manufactured in China.

8 7 6 5 4 3 2 1

CREDITS

We would like to thank the following vehicle owners and photographers for supplying the images in this book.

1940 Mercury: O: Richard Krist; P: Vince Manocchi. 1940 Pontiac: O: Larry Lange; P: Doug Mitchel. 1942 Buick: O: Dan Lempa; P: Doug Mitchel. 1942 Chrysler: O: James and Marian Humlong; P: Gary Smith. 1942 Chrysler: O: Paul Borgwardt; P: Vince Manocchi. 1942 DeSoto: O: James and Marian Humlong; P: Gary Smith. 1947 Chevrolet: O: Chuck Misetich; P: Vince Manocchi. 1947 DeSoto: O: Jim Schaffer; P: Doug Mitchel. 1948 Dodge: O: Dr. Roger Leir; P: Vince Manocchi. 1949 Hudson: O: W. E. Davis; P: W.C. Waymack. 1949 Plymouth: O: John Slusar; P: Thomas Clatch. 1950 Mercury: O: Ted S. Burleson; P: W.C. Waymack. 1953 Nash: O: Warren Danz; P: Doug Mitchel. 1953 Plymouth: O: Gary McIntosh; P: Vince Manocchi. 1954 Ford: O: Allan and Elaine Franklin; P: Vince Manocchi. 1954 Hudson: O: Edward E. Souers; P: Al Rogers. 1955 DeSoto: O: Tanya Johnson; P: Gary Greene. 1955 Dodge: O: Bob Riggs/Joel Johnson; P: Gary Smith. 1955 Mercury: O: Jimmie Clark; P: Jerry Heasley. 1955 Monarch: O: Dave Shelton; P: Vince Manocchi. 1955 Plymouth: O: Harry DeMenge; P: Vince Manocchi. 1956 Cadillac: RM Auctions; CM Media. 1956 Chevrolet: O: Richard Hibbard: P: Vince Manocchi. 1956 Pontiac: O: Andy Linsky; P: Phil Toy. 1957 Chevrolet: O: David Ertel; P: Tom Shaw. 1957 Peugeot: O: Dominique Legeai; P: Doug Mitchel. 1957 Ponrtiac: O: Andy Linsky; P: Phil Toy. 1958 Ford Consul: O: Ken Doehring; P: Doug Mitchel. 1958 Ford: O: Tom Turner; P: Vince Manocchi. 1958 Oldsmobile: O: John Strewe; P: Doug Mitchel. 1958 Simca: O: Mike Frankovich; P: David Cooley. 1960 Chevrolet: O: Eric P. Goodman; P: Lloyd Koenig. 1961 Plymouth: O: North Shore Classic Cars; P: Doug Mitchel. 1962 Chevrolet: O: Douglas Englin; P: Doug Mitchel. 1965 Ford: O: John and Connie Waugh; P: Vince Manocchi.

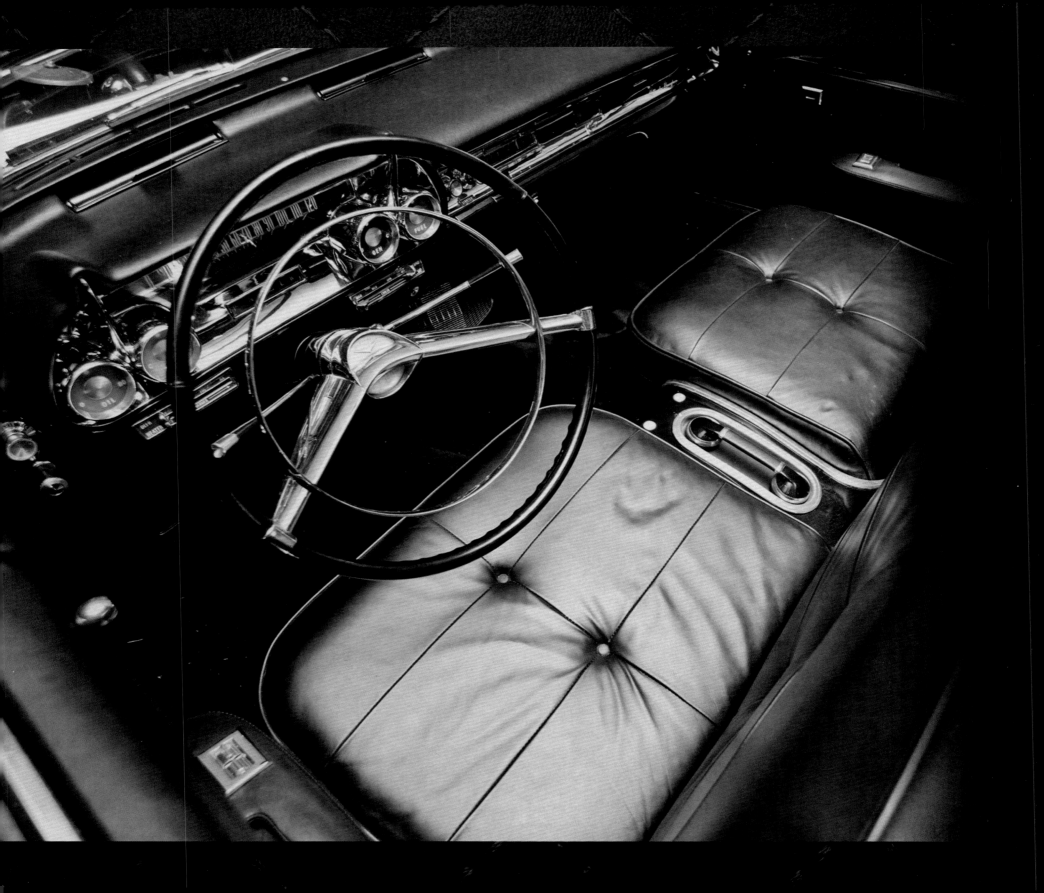

Table of Contents

1940 Mercury
Eight Convertible Coupe

In 1938, a gap of about $600 existed between the price of the Ford DeLuxe Fordor sedan and the company's Lincoln-Zephyr sedan. That was enough to buy a base Ford coupe. Ford was missing out on sales in the $800-$1400 market.

Edsel Ford had long wanted a medium-price car to fill the hole between Ford and Lincoln. While General Motors buyers could start with a Chevrolet and move up the Pontiac-Oldsmobile-Buick-LaSalle-Cadillac ladder, Ford customers had no place to go unless they became affluent enough to buy a Lincoln. That got a little easier with the introduction of the upper-medium-price Lincoln-Zephyr in 1936, but there was still a big void. Henry Ford—who called the shots—finally gave Edsel the go-ahead for such a car for the 1939 model year.

Edsel might have admired the GM "car for every purse and purpose" policy, but he didn't slavishly follow that formula. While GM tried to give each make a separate style and identity, Edsel planned a super-deluxe version of the Ford DeLuxe. The new car clearly shared styling and engineering with its Ford sibling. In fact, Edsel wanted the new car badged "Ford-Mercury." Chief stylist Bob Gregorie convinced Edsel at the last minute that the car should have a separate identity. Still, some early cars had "Ford-Mercury" on their hubcaps before "Mercury" hubcaps could be made.

Mercury's 116-inch wheelbase was four inches longer than Ford's, and sedan bodies were nearly eight inches wider. Mercurys were about 130 to 200 pounds heavier than similar Fords. Although styling was in the same vein, Ford and Mercury didn't share any body panels. There was more difference in interiors—Mercury was better trimmed and equipped.

Ford had developed a larger 239.4-cid version of its 221-cid V-8 for truck and police-car duty, and that engine was pressed into service for Mercury. The 95-bhp Merc engine had ten more horses than the 221. This gave the Mercury sedan a better power-to-weight ratio of 32.7 pounds per horsepower, versus 34.1 pounds for Ford. Ford had sprightly performance but the Merc was hotter still with a top speed of more than 90 mph. Mercury also got better gas mileage, thanks to a higher gear ratio. The successful new make ranked tenth in sales for its first year.

For 1940, there were styling tweaks and the badges were changed to "Mercury Eight." The gearshift lever moved from the floor to the steering column, and the convertible coupe gained a vacuum-powered top.

1940 Pontiac
Special Six Station Wagon

The car featured on these next few pages wasn't built by chance. Rather, it was the culmination of a series of events that swept through General Motors's Pontiac division in the late Thirties. Consider:

● "Silver Streak" styling. In 1935, Pontiacs appeared with a ribbed band of chrome down their centers. Designed in the Pontiac styling studio then headed by Franklin Hershey, the streaks would go on to serve as the make's visual signature for two decades. The same year, Pontiac returned to the production of six-cylinder cars with a 208-cid L-head engine good for 80 bhp.

● The arrival of station wagons. With demand for wood-bodied wagons growing steadily, Pontiac cataloged its first such car in 1937. It came in the DeLuxe Six series, in which the engine was newly enlarged to 222.7 cid and power was boosted to 85 bhp.

● A low-cost line. To cover a perceived price gap at the bottom of its lineup, Pontiac brought out the Quality Six for 1939. The 115-inch-wheelbase series used Chevrolet bodies and included a station wagon.

Come 1940, the Quality Six was renamed the Special Six, but the concept behind it was unchanged. GM created a new A-body for Chevrolet, but Pontiac adapted it for the Special Six, where updated Silver Streaks bisected a wide—for '40—grille of bright horizontal bars. Sealed-beam headlamps, which swept the industry that year, projected from the broad fenders.

Wheelbase grew by a couple of inches and there was more power under the hood. A compression boost to 6.5:1 (from 6.2) delivered 87 bhp at 3520 rpm. The Special Six was offered in five closed body styles, including Pontiac's lone station wagon. Prices ranged from $783 up to $1015 for the eight-passenger wagon.

The 1940 Pontiac Special Six Station Wagon had a wood body built by the Hercules Body Company of Evansville, Indiana.

1940 Pontiac Special Six Station Wagon

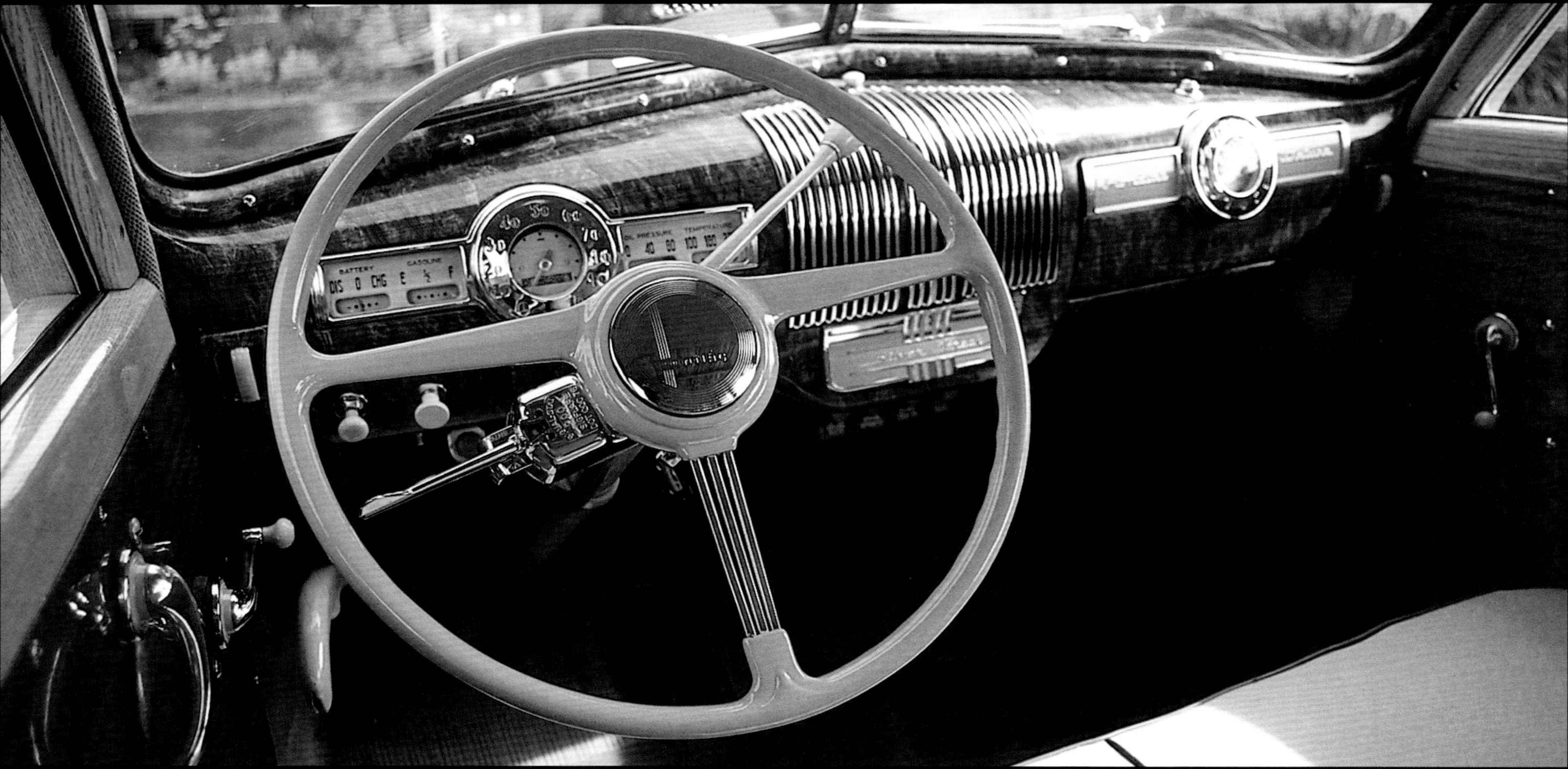

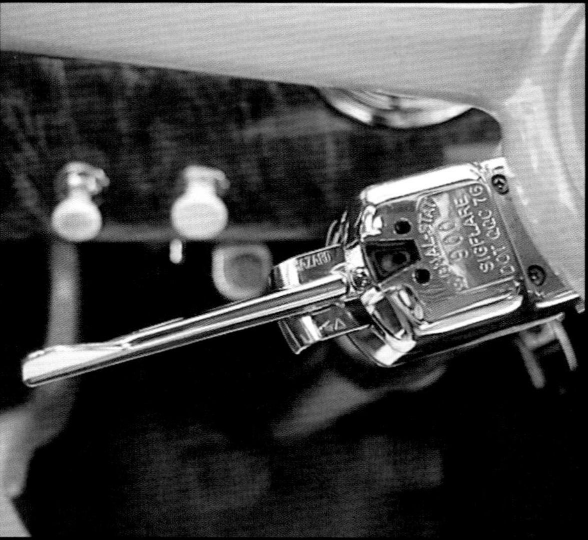

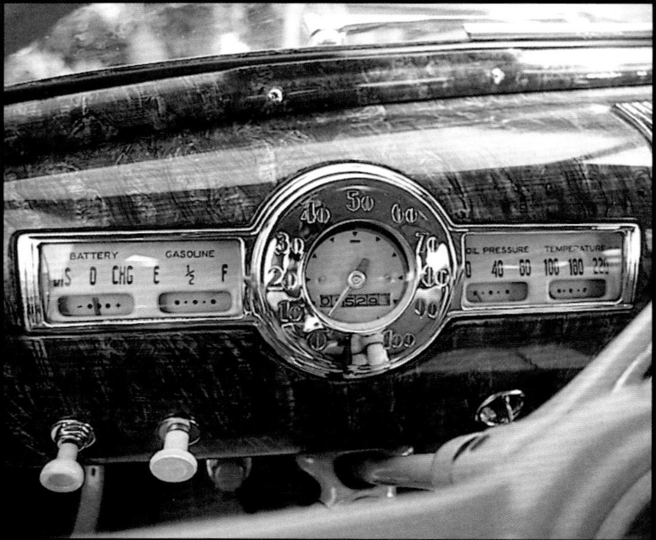

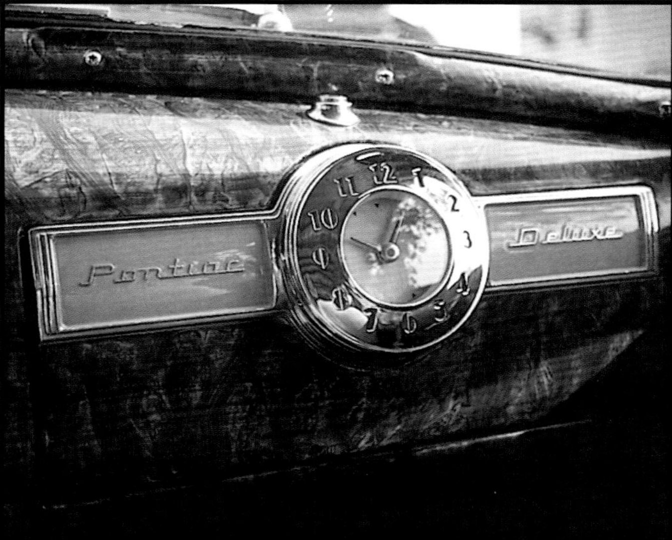

Special Station Wagon

This beautiful Buick Special Estate Wagon was General Motors's most luxurious wagon for 1942. Cadillac was content to leave such cars to other GM divisions and never offered a production station wagon until the CTS Sport Wagon of 2010-14. That left Buick the top dog of GM's wagon offerings from 1940 until the last Buick Century and Roadmaster station wagons in 1996.

Buick built a few depot hacks in the early Twenties but it wasn't until 1940 that Buick built its first real station wagon based on the midrange Super chassis. For 1941, the wagon moved to Buick's less expensive Special range. That doesn't mean that the Estate Wagon wasn't big and stylish. The Series 40 Special rode on a generous 121-inch wheelbase and shared styling with the other Buick series. Appearances were updated for '42 with front fenders that flowed into the doors and a distinctive wide grille with vertical teeth that would be a Buick trademark for many years.

It also shared an ohv straight-eight engine. Century, Roadmaster, and Limited had a 320.2-cid engine, while Special and Super used a smaller 248-cube version. For 1941, the Special engines developed 115 bhp with a single carburetor or 125 bhp with the available "Compound Carburetion" dual-carb system. In '42 those figures fell to 110 and 118, respectively, as cast-iron pistons were substituted for aluminum to free up the latter for defense-production needs.

The wood bodies were built by the Hercules Body Company of Evansville, Indiana. Hercules also supplied wagon bodies for the other General Motors divisions, as well as for Packard. Hercules used high-quality ash framing with mahogany panels.

This car is just one of 328 Estate Wagons made for the war-shortened 1942 model year. The wagon weighed 3925 pounds and cost $1450, making it the heaviest and most expensive car in the Special lineup. Oddly, a right-side brake light was optional. The tailgate-mounted brake lights were hinged so that they could be seen with the tailgate lowered. Buick was the first to have standard turn signals for 1939, and the rear turn signals were incorporated in the emblem below the tailgate. The turn-signal lever is mounted on the right side of the steering column, under the shifter.

Buick returned to the station wagon market with the 2018 Regal TourX wagon. The sleek TourX is sportier than the 1942 Estate Wagon, but it can't match the elegance of a Forties Buick "woody."

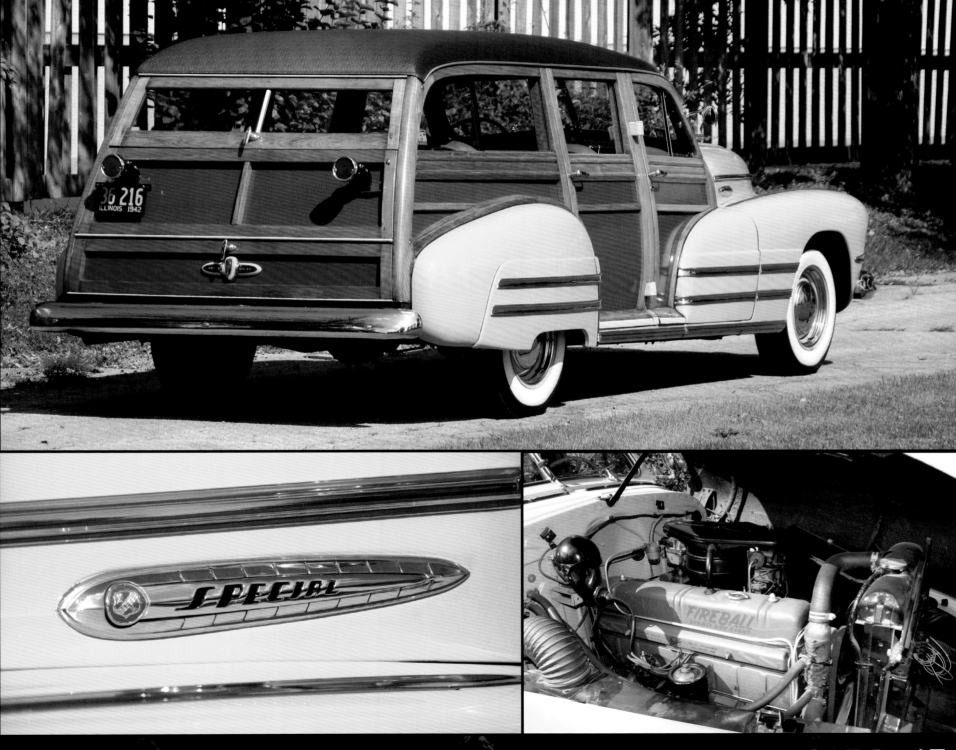

1942 Chrysler
New Yorker Four-Door Deluxe

Although war raged in Europe and Asia, America was at peace when the 1942 Chryslers were introduced in fall 1941. Americans hoped to avoid being drawn into the conflict and, perhaps optimistically, the '42 Chryslers used much of the soon-to-be-rationed chrome. The bombing of Pearl Harbor ended U.S. isolation and after December 31, 1941, chrome was diverted for war production. Cars made do with painted trim in place of the strategically important metal. No Chryslers were produced between January 29, 1942, and December 1945 as Chrysler and other automakers devoted themselves entirely to the war effort. Chrysler built artillery, marine engines, aircraft components, and 25,000 tanks during World War II.

To resume postwar car production as quickly as possible, the '42 Chryslers returned updated with a new grille and front fenders. The prewar design served well until a new Chrysler arrived in spring 1949.

This car was delivered in October 1941 and glimmers with a full complement of chrome. The Crown Imperial was Chrysler's flagship, but New Yorkers topped the "regular" line. Priced at $1495, this big sedan was smooth riding, roomy, and had a plush interior with wool broadcloth upholstery. Power was provided by a 323-cid 140-bhp L-head straight eight driving through a Fluid-Drive semiautomatic transmission. The combination didn't provide neck-snapping performance, but it was durable and dependable.

1942 Chrysler
Town & Country Station Wagon

By the late Thirties the utilitarian station wagon was developing a cachet similar to today's luxury SUV. However, upscale Chrysler didn't have one until 1941. To make up for the late start, Chrysler Division General Manager David Wallace planned a luxury wagon that was more streamlined and stylish than the competition.

While other wagons mounted a wooden box behind sleek front sheetmetal, Chrysler streamlined the entire vehicle. The rear was a semifastback or "barrelback" design. It was the first wagon with a steel top, a stamping shared with Chrysler's limousine. The new creation was badged Town & Country—a nameplate subsequently used almost continuously by Chrysler through 2016.

Chryslers of the period had bulletproof construction and the Town & Country was no exception. The ash frame and mahogany panels were milled by a Chrysler subsidiary. A section of Chrysler's Jefferson Avenue plant in Detroit was devoted to the laborious task of assembling the wood-and-steel body.

The Town & Country was mounted on Chrysler's 121.5-inch six-cylinder chassis. The 1941 Town & Country was part of the base Royal line, but moved up to the better-trimmed Windsor family for 1942. The 250.6-cid L-head six developed 120 bhp in '42. A semiautomatic Fluid Drive and Vacamatic transmission was standard.

Town & Country was available as a six-seater or as a nine-passenger version with three rows of seats. The middle seat folded similar to limo jump seats. The rear bench was hinged to either pivot back to provide legroom with the jump seat folded down, or forward to create more cargo space. Leather upholstery was standard.

The T&C accounted for about 2000 sales between 1941 and '42 before World War II ended production. The timbered Town & Country returned postwar, but in convertible, sedan, and coupe versions. The name finally returned to wagons in mid-1950.

Custom Coupe

With 1941 having been a big year for new body designs from most U.S. automakers, 1942 was destined to be a big year for styling facelifts. Typically, grilles widened, fenders stretched, and a little more chrome was added to cars in top trim levels.

DeSoto was right in step with the industry trend. To its year-old "Rocket" bodies it added some new touches that were common to other '42 Chrysler Corporation products, plus a few that were highly distinctive.

Perhaps the most talked-about feature on the new DeSoto was its "Airfoil" headlights. Squarish doors hid the lamps until an underdash lever rotated the covers upward and turned on the lights. Though hidden headlights had been used on 1936-37 Cords, they were still very much a novel idea at the time.

The hideaway headlamps were joined by a new take on the vertical-tooth grille first seen on 1941 DeSotos. In place of divided, winglike grille sections, the '42s sported a lower, wider, and more uniform array of chrome bars in a slight "S" shape. (While the Airfoil lights were used only on the war-shortened run of 1942 cars, the toothy grille was destined to become a regular DeSoto feature into the mid-Fifties.)

Like other Chrysler products, the '42 DeSoto came with doors and body panels that flared at the bottom to hide the running boards. Bumpers were enlarged and curved at the ends, and parking lights moved to a spot in the fenders near the grille.

There was one notable technical change for the '42 DeSoto. Its L-head six-cylinder engine was bored out to 236.6 cid, a gain of 8.5 cubic inches from '41. With the added displacement and larger valves, a 10-bhp gain—to 115—was realized. DeSoto continued to offer two coupe styles: a long-deck three-passenger type, and the six-passenger style with rear-quarter windows featured here. Both were available in DeLuxe and flossier Custom trim.

There was a special option package available for most Custom body styles. This was the Fifth Avenue group, which essentially bundled together all the available accessories. Features included fender skirts with ribbed bright trim, wheel trim rings, plaid Sportsman upholstery, radio, heater, dual horns, lighted hood ornament, and hoodside Fifth Avenue badging. Also optional was a Simplimatic four-speed semiautomatic transmission.

1947 Chevrolet
Fleetmaster Fleetline Aerosedan

Let's say you're not convinced that appearance is an important factor—maybe the important factor—that drives car shoppers to choose one vehicle over another. Then consider the 1947 Chevrolet Fleetline Aerosedan. Despite being the most expensive two-door closed car in the Chevy lineup, it was still the most popular model of the best-selling brand in America in '47.

Why? After all, mechanically, it was the same as any of Chevrolet's other ten models offered that year. In standard equipment, it varied little from most cars in the Fleetmaster series, out of which the two-car Fleetline subseries was derived. It had to be the looks.

Aerosedans had the benefit of fastback styling that was very much in vogue in the Forties. General Motors gave the style a big boost in 1941 when it launched a well-received armada of B-body Pontiac, Oldsmobile, Buick, and Cadillac sedans and coupes. When a fastback roofline was extended to the A-body program in 1942, Chevrolet jumped on the bandwagon with the Aerosedan.

An instant hit, this two-door companion to the notchback Sportmaster four-door sedan (another trendy style introduced in 1941 as the first Fleetline model) topped the Chevy sales charts in war-shortened '42. When postwar production resumed for 1946, the Aerosedan slipped back into the pack, but then shot back up into sales leadership the next two years. Our featured car is one of 159,407 Aeros made in 1947. Another 211,861 were made for '48 before the aging prewar design was finally replaced.

Aside from its distinctive roof, the Aerosedan came with a trio of chrome spears on each fender and special "Fleetweave" upholstery fabric that were Fleetline series exclusives. Otherwise, from its 90-bhp ohv 216-cid six to its 4.11:1 rear axle, the Aerosedan was a lot like most Fleetmaster models.

Custom Club Coupe

DeSoto reappeared after World War II looking fresher than most of its competitors. Like them, these Series S-II DeSotos were based on a model that had ceased production in early 1942. However, exterior styling was extensively updated, somewhat surprising since the short-lived 1942 models wore a facelift of their own.

The biggest changes were up front, where the hidden headlights that bowed for 1942 did not return. Thus, DeSotos once again sported traditional exposed lamps. The grille was altered too, as were the front fenders that now cleanly blended into the doors. A new front bumper wrapped further around the corners.

Out back, the rear fenders were reworked with squared-off rear wheel openings, and the restyled taillamps were larger and vertically oriented, in place of the horizontal units used before. The rear bumper was restyled along the same lines used up front, including the increased wraparound at the ends.

As they had before the hostilities, DeSotos came in DeLuxe and Custom trim on two wheelbases. "Standard" models rode a 121.5-inch span, while the extended-length seven-passenger models used a 139.5-inch spread.

Under the hood, DeSoto still relied on a 236.6-cid L-head six-cylinder engine, but it was now rated at 109 bhp, rather than the 115 it generated in the '42s. The standard transmission remained a three-speed manual.

The available semiautomatic transmission was tweaked a bit. The improved setup was labeled the Cyrol Fluid Drive with Tip-Toe Shift, no doubt by advertising types who were paid by the word.

Fluid Drive used an oil-filled housing that contained an impeller and a turbine. The impeller was powered by the engine, and it in turn moved the oil to spin the turbine that transmitted power to the clutch. The fluid eliminated the need to clutch while braking because the impeller and turbine were not mechanically linked.

The second half of the drive system was the Tip-Toe Shift, in effect a four-speed semiautomatic transmission. The driver only used the clutch when selecting reverse or shifting between the Lo or Hi drive ranges. Typical driving was handled in Hi, where the car started out in third gear. When the car had accelerated to about 15 mph, the driver let off the gas and the shift to fourth happened automatically. Downshifting also happened automatically while decelerating or flooring the gas pedal for a needed burst of acceleration.

The 1946-model DeSotos remained in production through early 1949 with few changes other than steadily rising retail prices. Model years for 1947 and '48 started on January 1 of those years; little other than the serial number gave a clue to a particular car's model year.

1948 Dodge
Derham Club Coupe

I f imitation is flattery, Studebaker must have been pleased by this one-of-a-kind Dodge coupe with a Starlight-like wraparound rear window. Dodge wasn't testing the waters for a glassy coupe model of its own but commissioned this special to move from dealer to dealer and generate showroom traffic.

Derham Body Company of Rosemont, Pennsylvania, did the conversion. Founded in 1887 to build carriages, Derham expanded to custom auto bodies in 1907. During the classic era, Derham was one of the top coachbuilders, its work gracing Duesenberg, Packard, and Pierce-Arrow chassis. As the super-luxury market dried up in the mid-Thirties, Derham survived by modifying production cars, performing handicapped-accessible conversions, and doing restoration work. While almost all American coachbuilders died in the Depression, Derham managed to survive until 1971.

In the late Thirties, Derham started building semicustom bodies for Chrysler and even acquired a Plymouth-DeSoto franchise. Thus, Derham was a natural choice when Dodge needed something distinctive to build showroom traffic.

The roof is what sets this Dodge apart, and the firm was an old hand at padded-roof installations. Easy-to-work wood was used to form the roof contours, but it was shielded from view and weather by fabric. The hubcaps are unique to the car and copy the trunk badge for their center emblems. The metallic-green paint was not a standard Dodge color.

Derham reworked the interior as well. The rear seat was moved back four inches for more legroom. The seats were upholstered in a special striped wool broadcloth with vinyl front bolsters. Plain broadcloth was used on the doors and sides. The dashboard was standard Dodge but was painted black as in convertibles instead of the woodgrain that was normal for coupes. A larger rearview mirror was used to take advantage of the panoramic backlight.

Mechanically, the car was typical '48 Dodge. A 230-cid inline six developed 105 bhp and was connected to a Fluid Drive transmission. Fluid Drive eliminated most shifting by combining a conventional clutch with a torque converter and electrical shifting circuits.

After the customized coupe's time on the dealer circuit, Dodge sold it to its first private owner in 1950.

Super Six Business Coupe

The postwar Hudson was radically different from its prewar predecessors. While some new postwar designs still rode on a prewar chassis, the 1948 Hudson was completely new.

Hudson moved from conventional body-on-frame to unitized construction, but what was really revolutionary about the new Hudson was that the floor was welded under the frame rails rather than above as on other cars at the time. Passengers stepped down when entering a Hudson, which is why its 1948-54 cars are commonly called "Step-downs." Hudson was able to market a car that was lower than the competition without sacrificing headroom or ground clearance. This not only helped appearance, but lowered the center of gravity, which was a boon to handling. Hudson dominated NASCAR racing in the early Fifties not because it had the most power, but because it could out-corner other cars.

The slab-sided styling was considered futuristic when introduced. Ads proclaimed, "Now ... you're face to face with Tomorrow!" And, for a few years, Hudsons were modern. The streamlined-torpedo trend of the prewar years reached its ultimate expression with the "bathtub" style of Hudson, Packard, and Nash in the late Forties. However, a more squared-off style appeared after the war and soon gained acceptance. Hudson's unibody construction made restyling more difficult and this would later hurt the company in the Fifties as its rounded styling became more out of step with the times.

The Step-down Hudson's straight eight was a prewar design, but the more popular six-cylinder engine was new. The 262-cid 121-bhp engine was the biggest, most powerful six in America. Although a fresh design, the six retained side valves even though the industry was moving to overhead valves. Still, Hudson would eventually get as much as 170 bhp from its six—comparable to the power of contemporary ohv V-8s.

The engine was connected to the three-speed manual transmission with an unusual "Fluid Cushioned" clutch, which was faced with cork inserts running in a bath of oil for smooth operation. Overdrive was an option that raised the cruising speed of this car to 70-75 mph.

At $2053, the Super Six business coupe was Hudson's least expensive car. Appealing to traveling salesmen, the coupe had a cargo deck instead of a back seat. Unusual for the time, a panel folded down to create a trunk pass-through.

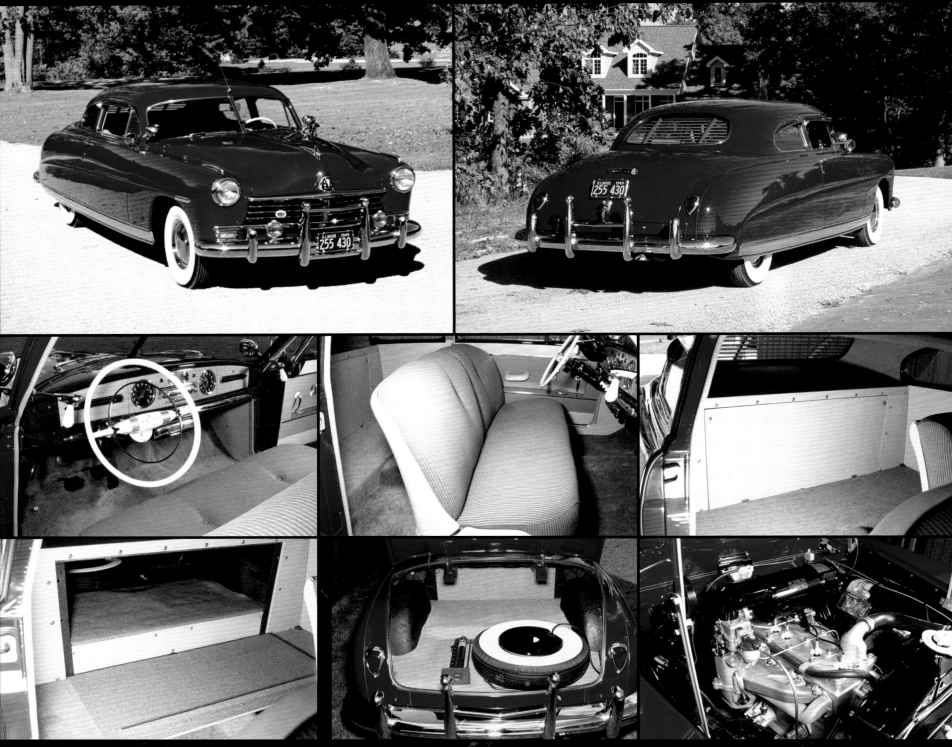

1949 Plymouth
Special DeLuxe Station Wagon

There's no arguing with the numbers. Even though Plymouth introduced its steel-bodied Suburban station wagon late in the 1949 model year, it sold 19,220 of them. Meanwhile, the traditional higher-maintenance wood-bodied wagon—available from the start of the selling season—drew just 3443 orders. Is it any wonder, then, that the "woody" was pruned from the Plymouth lineup after 1950?

It cost at least $2372 to secure the services of a '49 Special DeLuxe wagon. Suburban prices started at $532 less, but the difference bought more than 7.5 inches of additional wheelbase and two extra doors. The Special DeLuxe seated up to eight passengers with its third-row seat installed. Limited to two rows, the Suburban could hold no more than five. The Special DeLuxe also boasted nicer interior trimmings and more external brightwork.

After an extended run of 1948-style Plymouths, the "true" '49s arrived in the spring with fresh, but somewhat frumpy, styling. Even the U.S. Body and Forging wagon body was modernized, with a steel roof replacing the wood-and-fabric top of years past. The redesign also included a spare-tire hatch incorporated into the steel lower tailgate. A 218-cid "flathead" six returned under-hood, but with a two-horsepower boost to 97 bhp.

1949 Plymouth Special DeLuxe Station Wagon

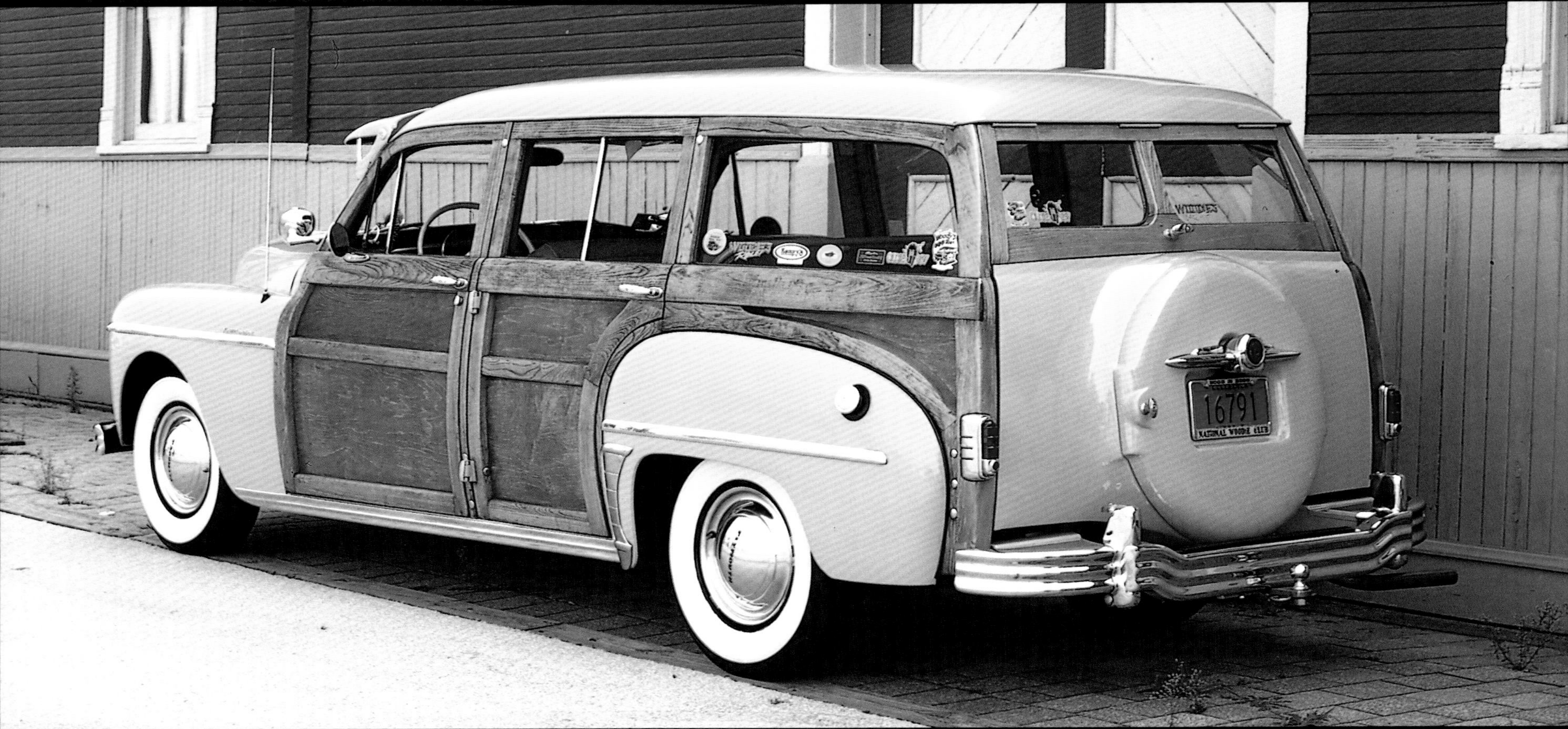

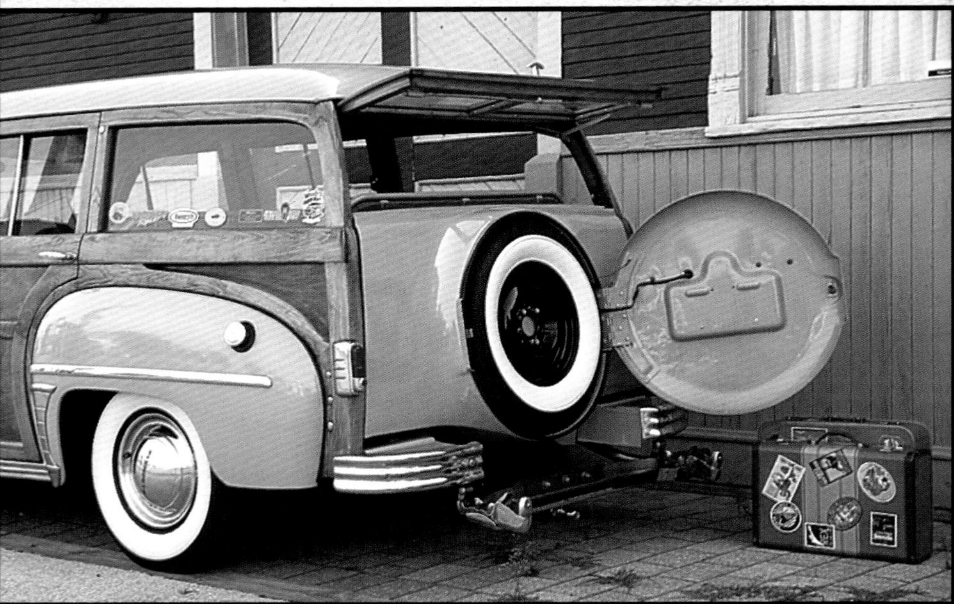

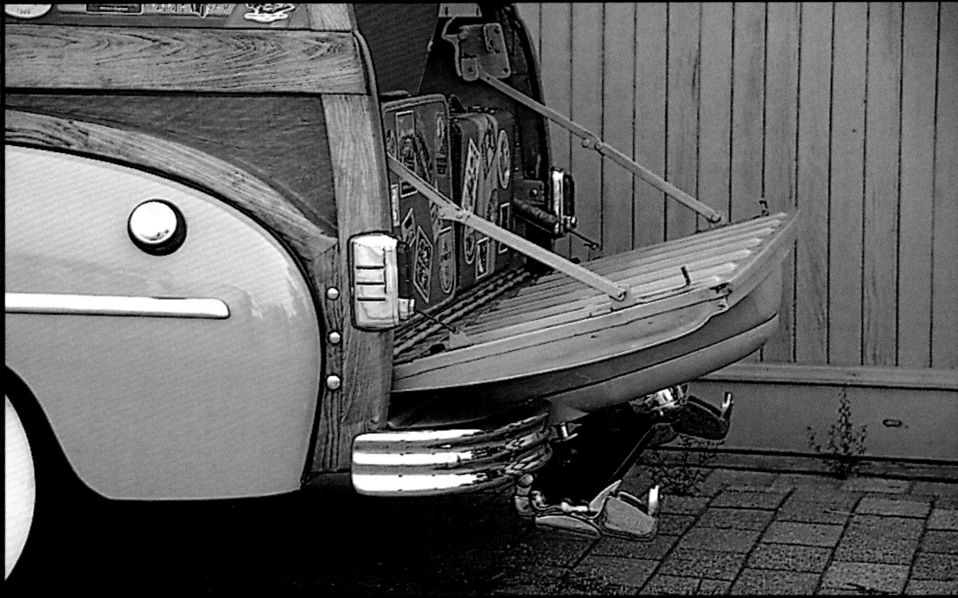

1950 Mercury
Monterey Coupe

In 1950 Ford Motor Company found itself suddenly surrounded by a legion of "hardtop convertibles" from rival General Motors and Chrysler Corporation marques, Ford determined it couldn't convert its year-old bodies into hardtops at a reasonable cost. But neither could it afford to stand idly by as GM and Chrysler wooed car buyers with their attractive new pillarless coupes. The temporary solution was to customize existing two-door bodies into a new line of specialty models: Ford Crestliner, Lincoln Lido and Capri, and Mercury Monterey.

All these cars were developed from the same formula. Grained top coverings harmonized with a limited number of paint colors, and color-coordinated interiors matched the external theme. Special trim and badging were used inside and out.

The Monterey's special touches were designed under the direction of William Schmidt, whose Lincoln-Mercury Division stylists performed a light facelift of Eugene Gregorie's 1949 Mercury design for 1950. The custom coupe came with a choice of canvas or vinyl roof coverings, and fabric-and-leather or all-leather seats. Bright trim edged the windows on the inside, and a deluxe steering wheel was used.

This Cortaro Red Metallic Monterey is topped by a black vinyl roof and features all-leather upholstery, options that together added $21 to the $2146 base price of a 1950 Monterey. Otherwise, it features the same chassis and running gear used on all other Mercurys. The 118-inch wheelbase runs between coil-and-wishbone suspension up front and leaf springs under a live axle in the rear. Power comes from an L-head V-8 displacing 255.4 cubic inches and making 110 bhp.

Regardless of the cost, Ford felt obliged to introduce its first hardtop in early 1951 to supplant the Crestliner. Mercury and Lincoln hung on with their gussied-up coupes until reengineered cars came out for 1952 with all-new bodies designed to include hardtops. The Monterey name continued, but on a premium series of Mercs in three body styles.

Rambler Custom Convertible

The Nash Rambler went against conventional economy-car wisdom when it bowed as a pricey convertible instead of a low-priced sedan. When the compact was introduced in 1950, World War II had been over for five years, yet raw materials were still regulated by the government and Nash wouldn't have been able to get enough steel to meet the expected demand for the new Rambler. Since production would be limited, Nash decided to build a high-profit car.

Besides increasing profits, the well-equipped convertible boosted Rambler's image. By contrast, Kaiser-Frazer's Henry J was introduced with a bare-bones "stripper" model that contributed to a "cheap-car" image that probably hurt sales.

The new Rambler followed Nash styling and engineering conventions. Nash was an early adopter of unitized construction that it labeled "Airflyte." Nash claimed that Airflyte construction reduced the Rambler's weight by 200 pounds and also made the car stiffer and less likely to squeak or rattle. Unlike other convertibles, the Rambler had fixed side roof rails with a power top that retracted on the rails. The side rails retained enough of the roof's rigidity that the kind of extra underbody bracing required in other convertibles was not needed.

Styling echoed that of the full-sized Nashes. The big Nashes were restyled for 1952 with input from Italian coachbuilder Pinin Farina, and Rambler's 1953 facelift again mimicked the senior Nashes. For marketing purposes, Farina was given credit for the design and Ramblers wore his trademark "F" logo on their flanks—even though the shop in Turin had little to do with the design.

The hood ornament was a purely American work. Chicagoan George Petty drew pinup girls for advertising and *Esquire* magazine. During World War II his art gained even more fame as it was often reproduced on the noses of military planes such as the Memphis Belle B-17 bomber. Nash recruited Petty to design voluptuous hood goddesses.

Under the Petty-girl hood ornament was a flathead six similar to the one in the Nash Statesman. Cars with manual transmissions had a 184-cid unit with 85 bhp, while cars equipped with optional Hydra-Matic automatics got a 195.6-cid 90-horse version. In the light 2590-pound Rambler convertible, the engines gave good performance and fuel economy for the time. In spite of its trim size, Rambler had room for five passengers.

A famous Rambler convertible driver was Lois Lane of the 1952-58 *Adventures of Superman* TV series. Lois started out in a 1951 Rambler convertible that was later traded in on a 1953 version similar to the example featured on these pages.

Nash sold 3284 Rambler Custom convertibles in 1953, but very few remain today. By 1954, Nash offered Ramblers in six body styles, including four-door sedans and wagons. Sales of the expensive convertible fell to around 200 units and it was dropped from the '55 line.

1953 Nash Rambler Custom Convertible

1953 Plymouth
Cranbrook Convertible Coupe

In 1953, the U.S. economy was robust. Bestowed with fresh styling, Plymouth set a record with almost 650,000 cars built while retaining its number-three sales position behind Chevrolet and Ford—as it had since 1931. That year was also Plymouth's 25th anniversary, but it chose not to celebrate. Perhaps with Ford and Buick celebrating golden anniversaries that year, Plymouth felt like an upstart.

The restyled Plymouth was smoother, rounder, and sported its first one-piece curved windshield. All Plymouths rode on a 114-inch wheelbase. At a time when the competition was getting bigger, Plymouth's overall length actually shrank an inch. However, Chrysler Corporation management was more concerned with function than form. Interior room—head room in particular—counted for more than a sleek silhouette. Although lower than the previous year's car, the new Plymouth was still tall and upright compared to the competition.

Ride was a functional concern, too, and Chrysler Engineering made sure that Plymouth's suspension was the best of the "low-priced three." The engine was a dependable 217.8-cid L-head six that dated back to 1942. With only 100 bhp, Plymouths were less powerful than Fords or Chevys, but they were also lighter, so performance was comparable. Plymouth was the only make of the three that didn't offer an automatic transmission option but at midyear a semiautomatic was added. Dubbed Hy-Drive, it replaced the flywheel with a torque converter and eliminated much (but not all) of the need for shifting and clutch work.

Plymouths came in two trim levels: base Cambridge and better-equipped Cranbrook. The $2220 convertible was the most expensive model, and was naturally part of the Cranbrook series. The convertible and all other two-door models had a new "E Z Exit" front seat with the seatback divided in a one-third/two-thirds split. This allowed a front passenger to slide over to the center of the seat while the outer third of the seatback was folded forward to allow rear passengers to get in or out.

After the record sales of '53, Plymouth production for 1954 fell to fewer than 463,000 cars, slipping to fifth place behind surprise third-place claimant Buick and fourth-place Oldsmobile. A concerted sales battle between Ford and General Motors, along with Plymouth's "practical" styling, worked against Chrysler's high-volume brand. But Plymouth fought back. By '55, it had fresh Virgil Exner "Forward Look" styling, available V-8 power, and a PowerFlite automatic transmission. Sales topped 700,000 that year. Two years later, Plymouth reclaimed third place.

1954 Ford
Crestline Sunliner Convertible

The 1954 Ford was a pivot point for its manufacturer. On one hand, it carried over the styling—albeit facelifted—the body styles, and the series names in place since 1952. On the other hand, the '54 ushered in certain engineering and detail changes that would live on in its successors.

Foremost among the changes was the switch from Ford's long-serving "flathead" V-8 to a completely modern short-stroke ohv mill. Dubbed the "Y-block" because of its deep crankcase, it displaced the same 239.4 cubic inches as the engine it replaced, but generated 130 bhp, 20 more than its forebear. It would grow larger and more powerful over the next few years.

This was also the year that ball-joint front suspension, already proven in Lincolns, was adopted to improve ride and handling. It was accompanied by a half-inch wheelbase extension. The resulting 115.5-inch wheelbase would serve Fords through 1956.

Then, too, 1954 was the year of the "Astra-Dial" speedometer, which admitted natural light through the rear of the speedometer housing to illuminate the dial. The feature would continue into 1955.

Another of the year's highlights was the first Skyliner, a Victoria two-door hardtop with a Plexiglass roof insert. This feature would be available for two more years. Other new models for 1954 included a two-door Ranch Wagon station wagon in the Customline series and a four-door sedan in the top-end Crestline series.

The hands-down glamour queen of the '54 Ford line was the Crestline Sunliner convertible, or so thought the customers who placed an impressive 36,685 orders for it. Prices started at $2164 for six-cylinder models, or $2241 when equipped with the V-8.

This Sunliner's V-8 drives through a Ford-O-Matic automatic transmission ($184). It's also festooned with the power assists then beginning to show up even on low-priced makes: window lifts ($102), brakes ($41), steering ($134), and front-seat adjuster ($64). Also adding to the Sunliner's final tab, were deluxe radio, heater/defroster, tinted glass, back-up lamps, and an accessory Coronado continental kit.

et-Liner Convertible Prototype

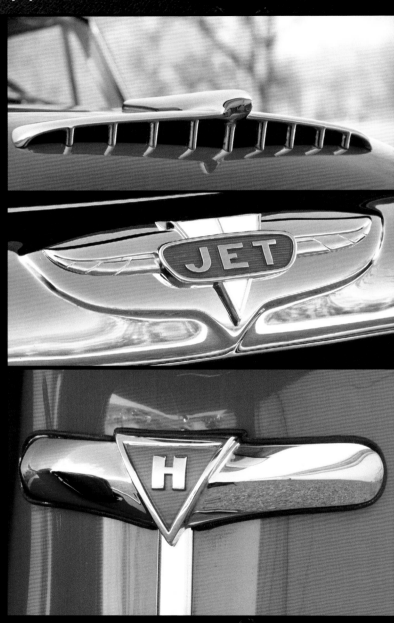

When it comes to collecting just about anything, rare is usually good. When it comes to cars, few are rarer than factory-built prototypes. Count among them the one-of-a-kind factory-built 1954 Hudson Jet-Liner convertible prototype.

Hudson bet big on the Jet, introducing two- and four-door-sedan versions of the compact for 1953. It rode a 105-inch wheelbase and was 180.7 inches long. The engine was a "new" 202-cid L-head six that was actually Hudson's old Commodore straight eight minus two cylinders.

Well engineered, solidly built, and a fine performer, the Jet suffered in the looks department. In an era of longer, lower, and wider, the small Hudson's tall and skinny shape was not a big selling point.

Price was a bigger problem. A stripped Jet four-door sedan started at $1858, or nearly $200 more than a basic Ford or Chevy. Worse, a top-line Chevy Bel Air four-door cost $1874.

With looks and price against it, Jet was a tough sell. First-year deliveries came to 21,143. Worse, total Hudson sales were down almost 10,000 units. Red ink flooded the ledgers, turning 1952's profit of about $8 million into a net loss of $10.4 million. A substantial portion of the losses are said to be attributable to adjustments that had to be made to the Jet's tooling amortization schedule.

Still, at some point, Hudson explored the possibility of expanding the Jet line in a bid to increase sales. The company likely was serious about adding a convertible, since this car was fully engineered with handmade convertible-specific body components, including a windshield lowered two inches from conventional sedan height. The prototype is trimmed as a Jet-Liner—a new top-end series added for '54—suggesting the model would have been priced at the highest level of the small Hudson's range. It was fitted with "Twin H-Power," Hudson's twin-carburetor option that helped the small six produce 114 bhp.

Hudson consolidated with Nash to form American Motors on May 1, 1954. Like the big "Step-down" Hudsons, the Jet went out of production that October. Had Hudson survived as an independent, a production Jet convertible likely wouldn't have made much of a difference. Rambler's compact ragtop was dropped after the 1954 model run, when only 221 had been built.

Firedome Convertible Coupe

Had Virgil Exner become a designer of buildings rather than cars, he would have been an advocate of Louis Sullivan's belief that "form follows function." Known for his innovative design work for Pontiac and the Raymond Loewy studio, Exner produced vehicles that showcased his approach to automotive design, "art made practical." One notable example of this philosophy is the 1955 "Forward Look" DeSoto.

Named for explorer Hernando DeSoto, the cars were supposed to appeal to the owner's sense of adventure. But uninspired, boxy postwar cars became the antithesis of this spirit. With sales of all its cars plunging, Chrysler had one option: invest more than $200 million in designing, developing, and tooling models restyled under Exner's direction. The investment paid off with record sales in '55.

This 1955 DeSoto Firedome is one of only 625 convertibles made that year in the marque's low-line series. Advertised as the car "styled for tomorrow," the new Firedome was lower, wider, and longer than before on a 126-inch wheelbase. A 291-cid "hemi" V-8 engine with a two-barrel carburetor made 185 bhp, enough to propel the Firedome to speeds up to 100 mph, all on regular gas. Even when resting, the Firedome appears to be in motion with its wraparound New Horizon windshield, aerodynamic hood ornament, and chrome-outlined contrasting color sweeps that were new for the year.

There was nothing shy about the Firedome, beginning with its grinning eight-toothed chrome grille at the end of its sloped, rounded nose. The red and white paint job on this car was one of 55 color combinations offered, each with a coordinated interior color scheme.

Enhancements to the interior were more than cosmetic. The gullwing-shaped dual-cockpit instrument panel included a glove compartment, radio, clock, courtesy lights, and the chrome-plated "Flite Control" shift lever for the PowerFlite transmission. Called the "finest completely automatic transmission available," the two-speed PowerFlite helped the driver navigate through traffic, rock the car out of snow or mud, brake down steep hills, and slip into tight parking spaces.

All five Series S-22 Firedome models were priced at less than $3200 without optional equipment such as power steering, brakes, windows, and radio antenna; four-way power front seat; and air conditioning.

1955 Dodge
Royal Sierra Custom Station Wagon

The Keller years at Chrysler were definitely over in 1955. K. T. Keller, corporate president since 1935, liked conservative styling with high rooflines that allowed passengers to keep their hats on. By 1954, America wanted sleek, flashy cars and Dodge sales were suffering.

Chrysler's new chief of design, Virgil Exner, was the anti-Keller. His styling was dramatic even by Fifties standards. In '55, he got to show his stuff with new styling for all Chrysler divisions. Responsibility for the Dodges fell to Maury Baldwin, one of Exner's young recruits. The '55 Dodge had what copywriters called "Flair Fashion" styling. Two- and even three-tone color schemes were used to make cars appear longer and lower. The interior wasn't neglected with two-tone upholstery and a full complement of round gauges grouped in front of the driver. The public loved it, and Dodge sales increased by more than 75 percent for the model year.

Styling wasn't the only thing the new Dodge had going for it. Dodge got a V-8 in '53, and for '55 the engine grew from 241 to 270 cid. Horsepower for V-8s ranged from 175 to 193. The fastest models could top 100 mph, with 0-60 in the 14-second range. Both were good for a medium-price Fifties sedan. Top-line Custom Royals got a 183-bhp "hemi." An available power pack with four-barrel carburetor and dual exhaust raised horsepower to 193. Royals (including Royal Sierra Custom wagons) got a polyhead V-8 with 175 bhp. An L-head six was standard in base Coronets. A PowerFlite two-speed automatic transmission was optional. The shifter was a small lever on the dash.

Dodge's suspension was praised by the press. *Motor Life* said, "The Dodge sticks to the road like a flying bug on a windshield." The mix of performance and handling paid off with a good showing in the '56 NASCAR season.

Not as exciting as the stock car racers but more practical were the Dodge station wagons. At the start of the model year, the Royal Sierra, available with two or three rows of seats, was the top Dodge wagon. Then, at midyear, a Royal Sierra Custom with extra trim was added. (The more expensive a Fifties Dodge, the longer the name.) Sierra Customs had chrome tailfins and a dipped beltline molding cribbed from the flashy Lancer models. A V-8 badge was added just below the dip. Not many Royal Sierra Customs were built.

1955 Dodge Royal Sierra Custom Station Wagon

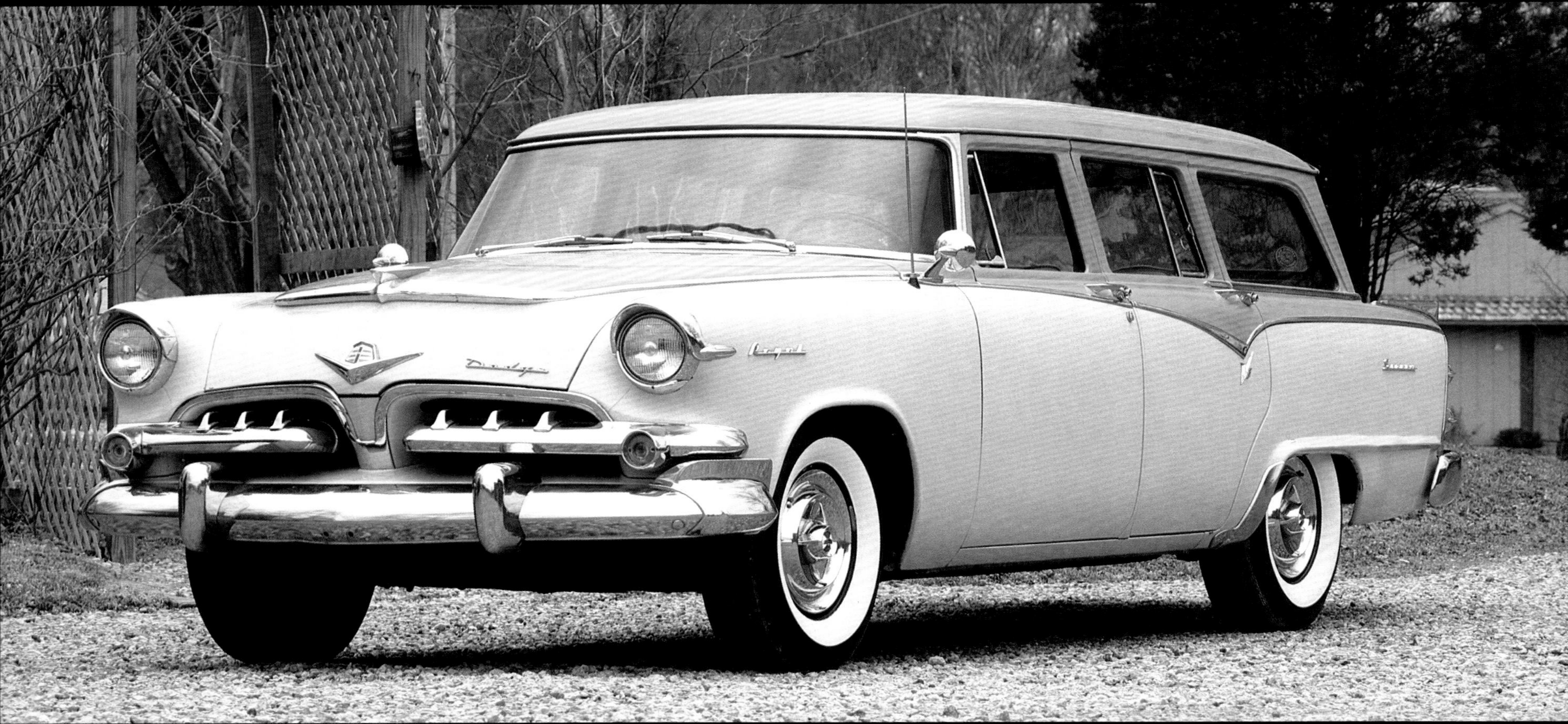

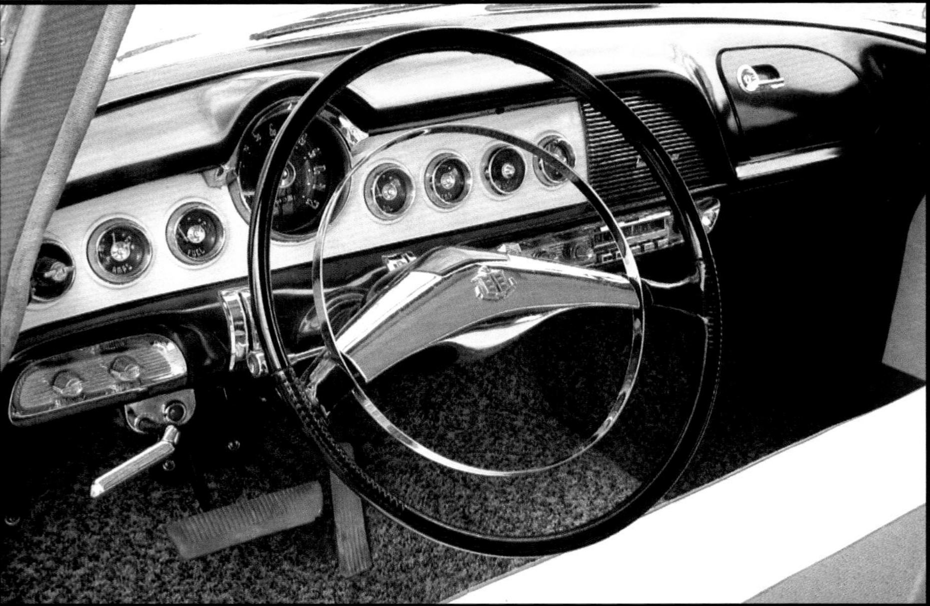

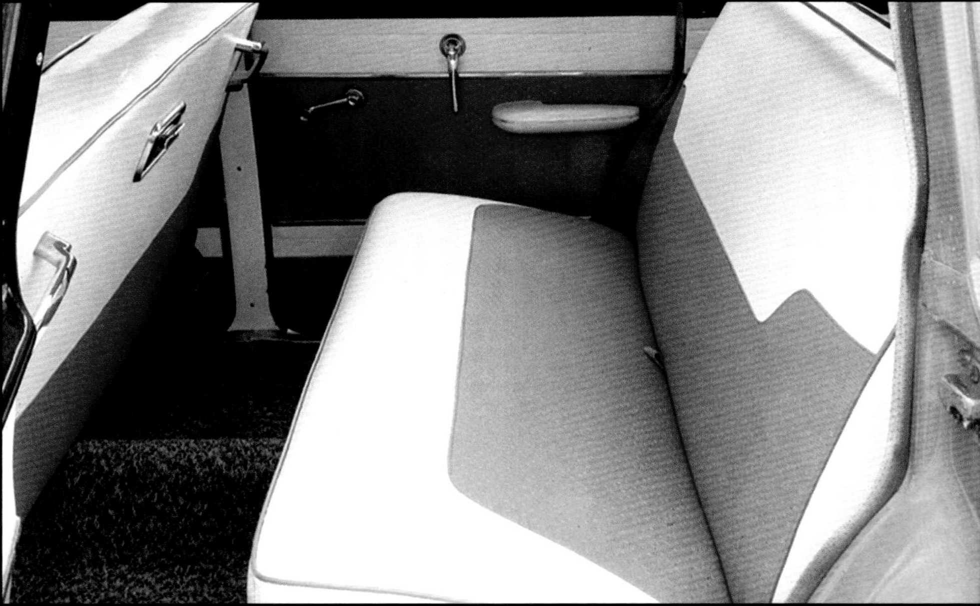

1955 Mercury
Custom Two-Door Sedan

While the new-for-1955 Montclair was the most stylish, most expensive Mercury, the Custom two-door sedan featured here was the lightest (3395 pounds) and least expensive ($2218) car offered by Mercury. The Montclair was priced to compete with the Buick Century, leaving the Custom to battle the popular Buick Special.

In April '55, Mercury split with Lincoln and became a separate division, but continued to share a bodyshell with Ford. Fresh styling included a new "Full-Scope" wraparound windshield. Fender skirts were optional on Customs, but standard on Montereys and Montclairs.

Inside, the new fan-shaped instrument cluster was rated by *Motor Trend* as "the most easily read of any '55 car." *MT* also found plenty of room in both the front and back seats.

The ohv V-8 introduced in '54 was enlarged to 292 cid. Refinements to the valvetrain and four-barrel carburetor made for a quieter engine and upped output from 161 to 188 bhp. Merc-O-Matic automatic transmissions gained an automatic low-gear start. Cars equipped with Merc-O-Matic could be ordered with a 198-bhp high-compression engine.

This extra power produced a good performer. *Motor Life's* pounds-per-horsepower ratings (based on gross weight) put Mercury at 21.0 pounds per pony while the rival Buick Special was rated at a less-advantageous 22.6. *Motor Trend* tested a Custom sedan with Merc-O-Matic and low-compression V-8, and considered it in the "hot car" class with a 0-60 mph time of 11.4 seconds.

Motor Trend also liked the handling, claiming "unexcelled roadability at high speeds and on curves." Chassis changes also gave a smoother ride.

The engineering and styling revisions helped Mercury hike model-year production to more than 329,000 units —its best year in the Fifties.

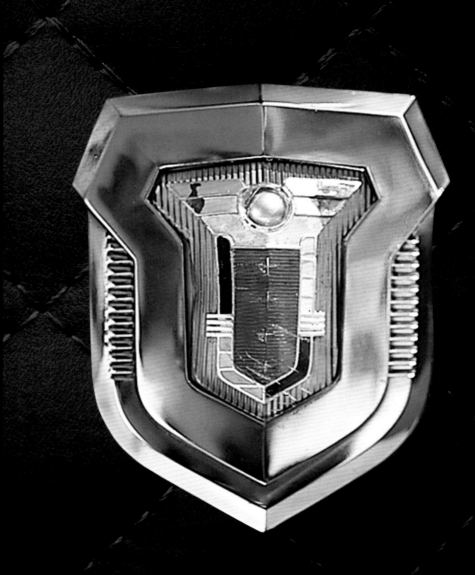

Richelieu Convertible

The population of Canada in the Fifties was less than ten percent that of the United States, yet Ford Motor Company sold up to six brands of cars in Canada. For as odd of an idea as that sounds, there was a reason for it.

Small Mercury dealers needed a lower-priced car to prosper. For them, the Meteor was created—essentially a Ford with unique trim. Meanwhile, Canadian Ford dealers wanted the ability to move customers up to a more expensive car. They could take on a franchise for the Monarch, a Mercury with trim and price variations. (Lincoln sold in small numbers and didn't get a Canadian alter ego.) Other American makes offered special models or variations of their brands for the Canadian market, but only Ford launched Canada-only nameplates.

Monarchs were built from 1946 to 1961, although they skipped 1958 to make room for Edsel. (That year was a recession year noted for garish styling that other makes might have liked to skip too.) Monarch outsold Mercury in Canada for much of its lifetime.

When Mercury was restyled for 1955 with more Lincoln-like appearance, per custom Monarch shared its new sheetmetal too. That same year, Mercury gained a new line-topping Montclair series—probably named for an upscale New Jersey suburb. Monarch's corresponding Richelieu took its name from a tributary of the Saint Lawrence River. The Richelieu differed from the Montclair with a unique front bumper, grille, hood ornament, side trim, and badging. Both makes listed a convertible only in these top-trim series.

The Mercury/Monarch engine was a 292-cid V-8 that developed 188 bhp with a 7.6:1 compression ratio or 198 bhp with 8.5:1 compression. The more powerful engine was standard on Montclairs with automatic transmission, but it was optional on Richelieus with an automatic. *Mechanix Illustrated*'s Tom McCahill tested a Montclair hardtop and sprinted from 0 to 60 mph in 12.8 seconds. He achieved a top speed of 106-108 mph. Similar performance could be expected from a Richelieu convertible. Mercury/Monarch had improved suspension for 1955, and *Motor Trend* proclaimed Mercury one of the best handling cars of that year.

Monarch had its best year with the stylish and powerful '55s. With 9623 units built for the calendar year, Monarch outsold Mercury in Canada. However, perhaps only 163 convertibles were built that year and it's estimated that only five remain.

1955 Plymouth
Belvedere Convertible

A "perfect storm" of styling pizzazz and V-8-engine performance swept through the American auto industry in 1955. The result was an industry record for the year in which more than 7.1 million of the 1955-model-year cars were sold. One of the greatest individual beneficiaries of this amped-up market was Plymouth.

Having slipped from its traditional third-place perch in sales to fourth behind Buick in 1954, Chrysler Corporation's low-price make bounced back handsomely in '55, turning out 704,454 cars. That wasn't enough for a return to its expected sales position (that would take another couple of years to accomplish), but it did mark a 52-percent gain from the year before and a slight increase in market share to back near ten percent.

The stumpy, compacted look that made Plymouth appear out of step with "longer, lower, wider" tastes in the 1953-54 period was neatly swept away for 1955. Virgil Exner's "Forward Look" set a new styling direction for all of Chrysler's brands. As executed under the direction of Plymouth chief stylist Maury Baldwin, the '55s stretched for an appearance of length, with front fenders that angled forward as they rose to their peaks and rear quarters that did likewise in the opposite direction. Only a grooved section at the grille center and triangular taillight lenses lent some visual connection to the model that came before.

Two new ohv V-8 engines provided lively alternatives to the carried-over L-head six: a 241-cid job good for 157 bhp, and a 260-cube powerplant that made 167 bhp with a two-barrel carburetor or 177 with a four-pot carb. PowerFlite automatic transmission—with a dashboard-mounted control lever—was available in place of the standard three-speed manual gearbox.

The new Plymouths rode a chassis stretched an inch in wheelbase to 115 inches and upgraded with alterations to the front shock absorbers and rear springs. Suspended pedals and tubeless tires were additional new features.

As had been the case in 1954, base Plaza, intermediate Savoy, and high-line Belvedere series were offered. The Belvedere was the best stocked, offering two- and four-door sedans, a two-door hardtop, and a Suburban station wagon with six-cylinder or V-8 power, plus a V-8-only convertible. The convertible's $2351 starting price was bested only by the tab for the V-8 Belvedere Suburban, but that couldn't keep 8473 sun lovers from motoring off from their local Plymouth dealership with one of the ragtops.

This Belvedere convertible has a 260-cid V-8 teamed with the PowerFlite trans. Other options include a radio and power steering.

CARS

Eldorado Brougham Town Car

The General Motors Motoramas of 1949-61 were car show extravaganzas designed to not only show off the latest of GM's offerings, but also test public reaction to dream cars. The Eldorado Town Car was part of a progression of dream cars that eventually led to a production car—the 1957-58 Cadillac Eldorado Brougham.

GM styling czar, Harley Earl, was the driving force behind Motoramas. Earl toyed with the idea of an ultra-luxury halo car for Cadillac. It started with the 1953 Orleans and the 1954 Park Avenue show cars. By 1955, the Cadillac Eldorado Brougham concept car was close to what would be the production car. Then, for the 1956 Motorama there were two Eldorado show cars. The Eldorado Brougham was a hardtop sedan that would see production the next year. The Eldorado Brougham Town Car concept was added to bring extra attention to the upcoming Eldorado Brougham sedan. The Town Car was distinguished from the sedan with a half roof over rear passenger compartment and an open chauffeur's compartment. There was a divider window between the front and back seats. The style recalled open-front limousines that were popular before World War II.

As was the case with many dream cars, the Town Car didn't have an engine and had to be pushed onto the stage. Its fiberglass body was easier to form than steel. Per limousine tradition, the chauffeur's compartment was upholstered in black leather. The rear compartment was less somber with beige leather and gold trim. Dual glove compartments for the back seat contained a decanter and cups, vanity case, and tissue dispenser. Telephones were provided for communication between the two compartments. The rakish Town Car was 3.5 inches lower than a standard Cadillac of the time. While the Brougham sedan had a brushed stainless steel top, the Town Car had a more formal padded leather top. The Town Car also had more restrained side trim than the sedan—although restrained is matter of degree. The Eldorado Brougham was the ultimate expression of Fifties flamboyant, jet-fighter-inspired styling.

The production Cadillac Eldorado Brougham cost $13,074, at time when Cadillacs started at $4677 and the most expensive limousine was $7678. Every conceivable option of the time was standard and included air conditioning, power front seat with memory, automatic headlight dimming, magnetized drink tumblers in the glovebox, a vanity with Arpège Extrait de Lanvin perfume, and air suspension. Power was provided by a 325-horsepower, 365-cid V-8 with dual four-barrel carburetors in 1957. For '58, the dual quad carbs were replaced by triple two-barrel carburetors and horsepower rose to 335. Only 400 Eldorado Broughams were sold in 1957 and 304 were built in 1958. A restyled Eldorado Brougham sold during 1959-60, but sales were even fewer.

As often was the fate of Fifties dream cars, the Eldorado Brougham Town Car was sent to a wrecking yard to be scrapped. Fortunately, the salvage yard didn't destroy the Town Car. In the late Eighties the car was rescued and eventually restored. During restoration, a '56 Cadillac engine was installed. Although not considered roadworthy, the Town Car can now move under its own power.

One-Fifty Two-Door Utility Sedan

The Chevrolets of 1955-57 have long been darlings among car enthusiasts for their iconic styling and legendary small-block V-8s. As is typical of most Fifties and Sixties collector cars, the most sought-after models are top-line, heavily optioned convertibles and two-door hardtops. Today, most Chevrolet collectors fawn over glitzy, two-toned Bel-Airs festooned with plenty of flashy factory accessories.

However, not every new Chevrolet shopper in the mid Fifties was swayed by fender skirts, bumper guards, tissue dispensers, signal-seeking radios, or Autronic Eye headlamp control. Some buyers just wanted to go as fast as possible for as little money as necessary, and were willing to forego creature comforts in the pursuit of speed. The age-old recipe for going fast on the cheap hasn't changed much over the years: Simply drop the hottest engine into the lightest, cheapest body available, and hold the frills. The 1956 Chevy shown here is a textbook example of the perfect budget bomb.

With a 3117-pound curb weight and a $1734 base price, the One-Fifty two-door utility sedan was the lightest and least-expensive model Chevrolet offered in 1956. A single chrome side spear (a new embellishment for the series) and small hubcaps highlighted its austere exterior trim. Interior accoutrements were equally Spartan: rubber floor mats; a single sun visor; and no radio, heater, or clock.

The rear windows were fixed—who needs roll-down rear windows with no back seat? The utility sedan featured a 31-cubic-foot cargo area lined by "durable composition-board loadspace walls" behind the single split-bench seat. While most '56 Chevys were available in an array of interior color choices, One-Fifty sedan buyers had no options; a beige and gold-striped vinyl with golden-flecked black pattern cloth was the sole trim combination.

The Chevrolet small-block V-8 was already well on its way to becoming a performance legend when the '56 models were introduced. A disguised 1956 Chevy prepared and driven by Corvette engineering guru Zora Arkus-Duntov set a new Pikes Peak hill-climb record on Labor Day 1955, and hot-rodders were quickly discovering how well the new engine responded to performance modifications.

Chevrolet engineers did a little hot rodding of their own when the Corvette dual-four-barrel 265 was made available as an across-the-board option midway through the 1956 season. With the first of the famous "Duntov" cams, dual four-barrel carbs, lightweight valves, and larger intake and exhaust ports, it was good for 225 bhp at 5200 rpm. The Carter carburetors were topped with a large "batwing" air cleaner from which hung two oil-bath air filters.

The options on this One-Fifty utility sedan are limited to the hot engine, Powerglide automatic transmission, and factory-accessory dual exhaust tips.

1956 Pontiac
Star Chief Custom Safari Station Wagon

The first Pontiac Safari was destined to forever be overshadowed by its Chevrolet Nomad relative. In total 1955-57 production, the Nomad more than doubled the Safari, 22,375 units to 9094. In the collector market, the Nomad is better known and brings higher prices at auctions. Yet Safari is an interesting variation on the Nomad theme that deserves all the recognition it can get.

General Motors's high-style wagons started as a Motorama show car. For the 1954 Motorama, GM styling chief Harley Earl asked for further body styles based on the new Chevrolet Corvette roadster. One concept was a station wagon dubbed Nomad.

The show car proved popular at the Motorama, and Earl pushed to have it put in production. Instead of a Corvette wagon, the production Nomad was part of the full-size Chevrolet line for '55. Corvettes were selling in small numbers; the two-door Nomad had a greater chance of success as a high-style Bel Air-series offering.

To increase the wagon's chance of profitability, it was suggested that sharing body-production expenses with a Pontiac version would lead to greater volume and, therefore, to lower costs. That spawned the Safari.

The Nomad and Safari shared a roof, tailgate, and glass. Even though Chevrolet and Pontiac used GM's "A" body, sheetmetal below the beltline was unique for each brand. Safari rode on a Pontiac chassis with a 122-inch wheelbase that was seven inches longer than the Nomad's span. The bigger Safari was 350 pounds heavier than a V-8 Nomad—which was also available with a six. To move the extra weight, Safari had a bigger V-8 engine. In 1956, the Pontiac V-8 grew to 316.6 cid while Chevy held at 265.

Reportedly, only 1056 Safaris were equipped with the standard manual transmission; the rest had a Hydra-Matic automatic. An advantage for Safari was that the Hydra-Matic had four speeds, while Nomad's Powerglide was a two-speed. With the automatic, the standard Safari V-8 developed 227 bhp. About 200 Pontiacs were equipped with a race-ready 285-bhp engine and manual transmission, but it's not known how many, if any, were installed in Safaris. The V-8/automatic Nomads had 170 or 205 bhp. (Perhaps 20 had a 225-bhp version of the 265 V-8.)

Safari was in Pontiac's top Star Chief Custom range and its interior was far more luxurious than Nomad's. Leather upholstery was a no-cost option, and a carpeted cargo floor was highlighted with stainless-steel strips. Nomad had nylon-and-vinyl upholstery, and its cargo floor was covered with linoleum.

Safari was highly regarded as a stylish twist on the utilitarian station wagon. It was low, with hardtoplike styling. Its chrome-ribbed tailgate had a rakish slant. At $3129, Safari was the most expensive 1956 Pontiac. Plus, it was a two-door wagon and less convenient than four-door models. The two-door Safari was dropped after 1957, but the name lived on with conventional Pontiac wagons.

Two-Ten Townsman Station Wagon

For as common as station wagons once were on the American scene, few of their names stand the test of time and memory. A rare exception is Chevrolet's Nomad of 1955-57, decked out in fancy trim and a rakish two-door roofline. Who, though, readily recalls the Handyman, the Beauville, or the Townsman—all of them sold in the same showrooms as the Nomad?

In the 1955-57 period, Townsman was Chevrolet's name for a four-door, six-passenger station wagon. (The nine-seat four-door was dubbed Beauville, and Handyman was the monicker for humble two-door, six-passenger wagons that weren't Nomads.) There was a Townsman in the top Bel Air series and in the midlevel Two-Ten lineup. The latter proved to be the most popular Chevy wagon of the year with 127,803 produced.

Though external sheetmetal and the instrument panel were completely new, the basic bodyshell and 115-inch-wheelbase chassis were little-changed from 1955. In appearance, interior decor of Two-Ten station wagons matched that of comparable sedans. However, where the sedans had cloth-and-vinyl upholstery, wagon seats were swathed solely in vinyl. Color combinations for wagons were ivory and charcoal, two-tone green, or beige and copper. Ribbed linoleum covered the cargo floor.

The base price for a '57 Two-Ten Townsman was $2456—with the standard six-cylinder engine and three-speed manual transmission. That total started increasing with the 283-cid V-8 engine and Power-glide two-speed automatic transmission found in this car. With its two-barrel carburetor and 8.5:1 compression ratio, this newly enlarged "small-block" mill made 185 bhp, but as much as 283 bhp was on tap with optional fuel injection. Power steering, an AM radio, and Sierra Gold and Adobe Beige paint are other extra-cost features that help make this Townsman memorable.

1957 Chevrolet Two-Ten Townsman Station Wagon

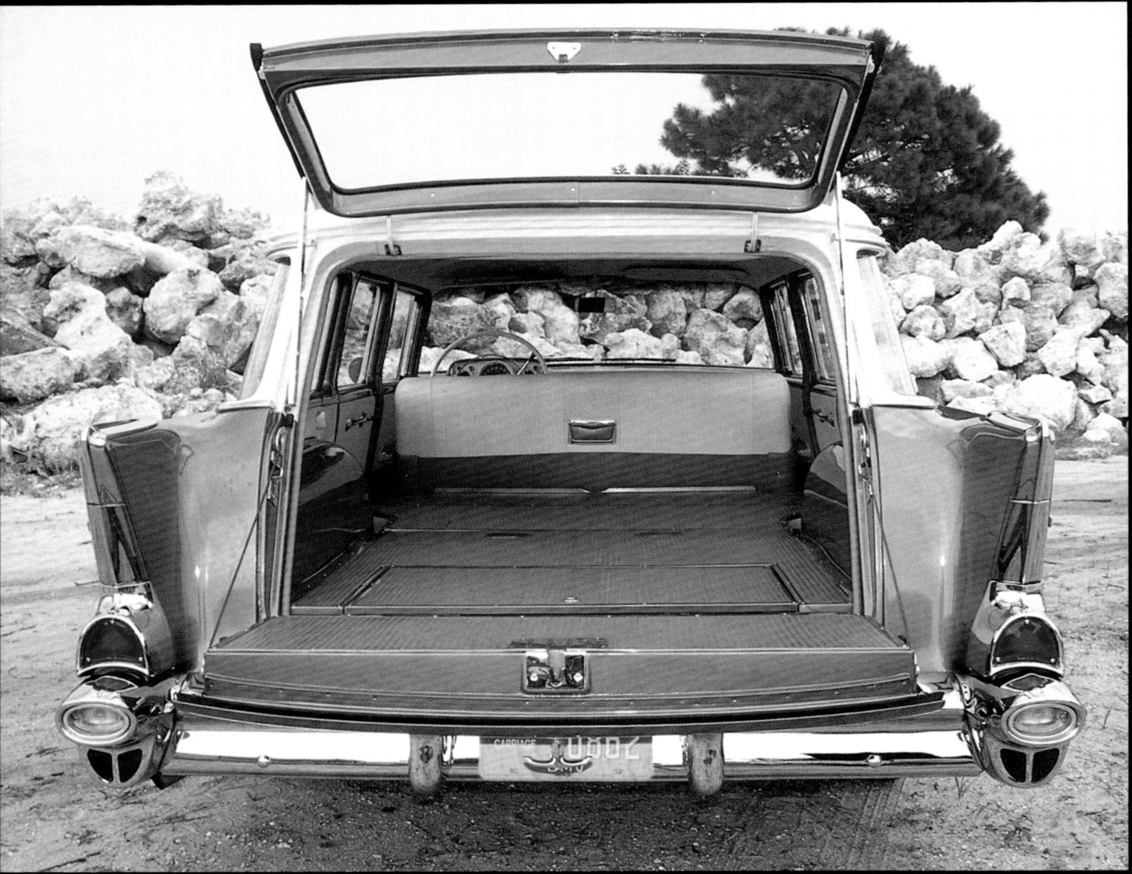

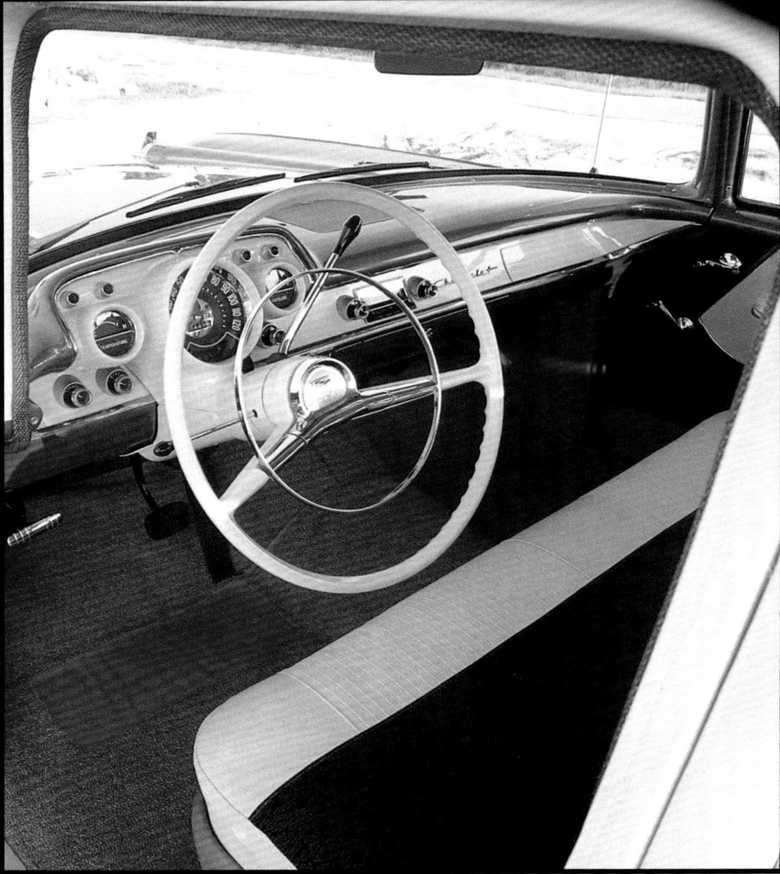

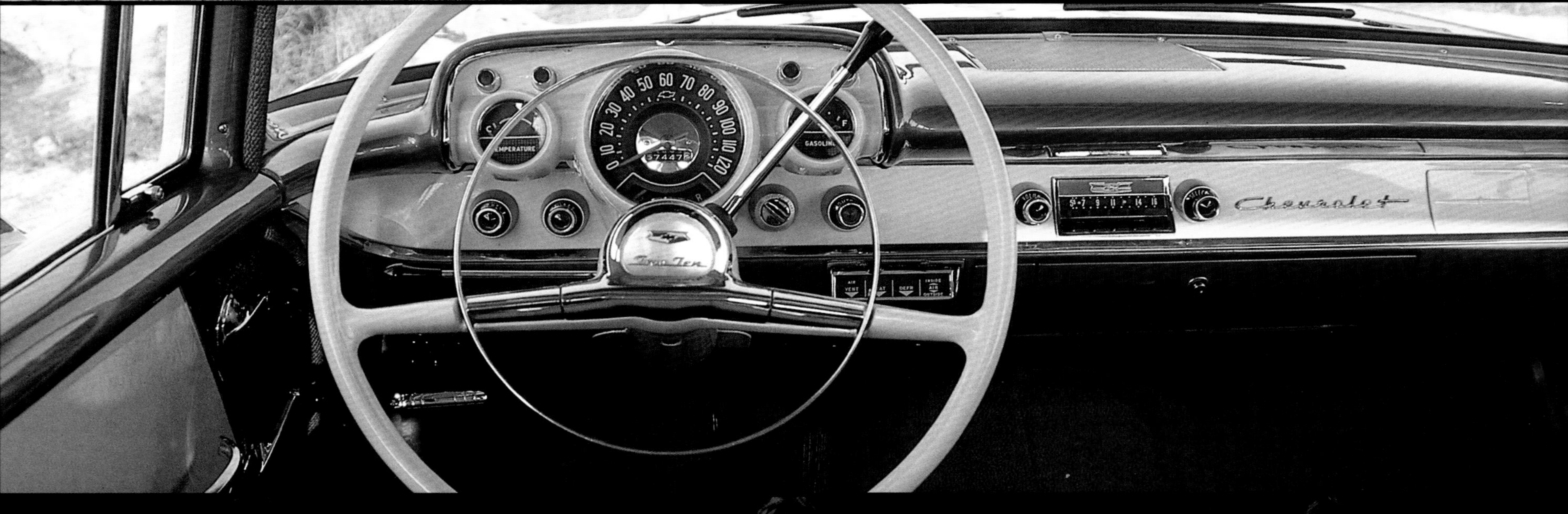

1957 Peugeot
03C Four-Door Sedan

ince their inception in France in 1889, Peugeot automobiles have fanned out around the globe. For a time, one of the places they could be found as the United States, where a Peugeot even became part of popular culture. On the television series *olumbo*, the rumpled police detective title character ove a tatty 403 convertible.

he 203C sedan featured here draws a blank with mericans, even those familiar with European cars, t's understandable. It dates from the year before ugeots were first formally imported to the U.S. (though all numbers of them had trickled in prior to 1958).

e 203 was Peugeot's first new design after World War First shown in 1948, it became Peugeot's sole model 949 after the prewar-style 202 was discontinued. In ril 1955, it was joined by the more modern 403 series, t 203 production continued until early 1960. By t time, nearly 700,000 had been built.

e unit-bodied 203 saw few changes to its basic design. ding had something of a Forties Chrysler look to it. rome bumpers were added in 1952. In 1954, the rear dow was enlarged and the fuel filler was hidden in right rear fender. Power came from a 1.3-liter ohv r with hemispherical combustion chambers that s good for about 42 bhp. The engine was hooked to our-speed manual transmission with an overdrive gear.

Star Chief Hardtop Coupe

The 1957 Pontiac was the first Pontiac since 1934 without "Silver Streaks" on the hood. Before Cadillac had tailfins and Buick had portholes, Pontiac had its signature chrome band (or bands) adorning the hood. However, Semon E. "Bunkie" Knudsen arrived as Pontiac's new general manager just as the '57 models were ready for production, and he was determined to change that.

Pontiac's position in the industry had been slipping. Knudsen believed a fresher, more youthful image was needed. Removing the Silver Steaks was seen as a break with the past—and a tradition that started while Knudsen's father, William S. Knudsen, was running Pontiac. It was also about the only change he could reasonably make at so late a date.

Actually, Pontiac's transformation from an "old man's car" had started in 1955. Pontiac shared Chevrolet's "A" body, which was new for '55. That year's Pontiacs were lower, sleeker, and more modern than before. Just as Chevrolet had a new ohv V-8 for '55, so did Pontiac. In fact, the Chevy engine's effective valve gear with lightweight, stamped rocker arms was borrowed from designs developed in Pontiac's engineering department. Pontiac's V-8 proved long-lived and versatile. The basic design lasted until Pontiac stopped building its own V-8s in 1981. While Chevrolet would eventually build "small-block" and "big-block" V-8s, Pontiac's sole V-8 design supported displacements from 287 up to 455 cubic inches.

The new V-8 cars had far more performance than the old straight-eight Pontiacs. To demonstrate the newfound power, Bonneville Salt Flats legend Ab Jenkins was recruited to set a world record. In his hands, a 1956 Pontiac averaged 118.375 mph for 24 hours—the last of many records set by Jenkins, who died later that year at age 73.

Bunkie Knudsen wanted to foster a performance image for Pontiac. Advertising in 1957 proclaimed Pontiac "America's Number 1 Road Car." Backing up that claim, Pontiac had increased the size of its V-8 to 347 cid, which helped it develop from 227 bhp up to 317 bhp. Styling was also more exciting for 1957. However, in spite of a good product, Pontiac sales were down for '57. Knudsen's efforts wouldn't truly start to bear fruit until 1959 when the "Wide Track" Pontiac rose to number four in sales.

Star Chief was Pontiac's top trim and offered a choice of cloth-and-leather or—as in this car—full-leather upholstery. The 1957 Pontiac Star Chief Custom Catalina hardtop coupe had a base price of $2901 and 32,862 were built.

1957 Pontiac Star Chief Hardtop Coupe

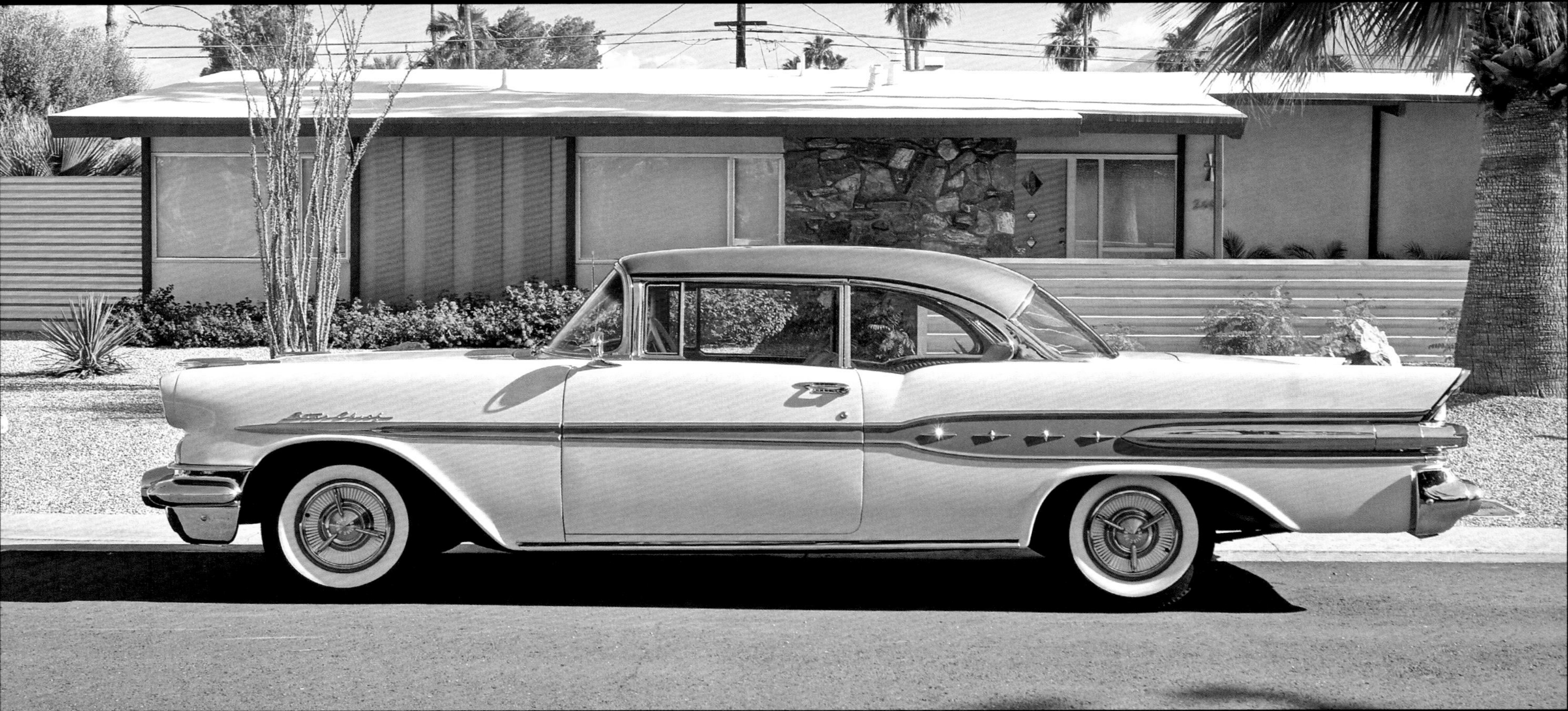

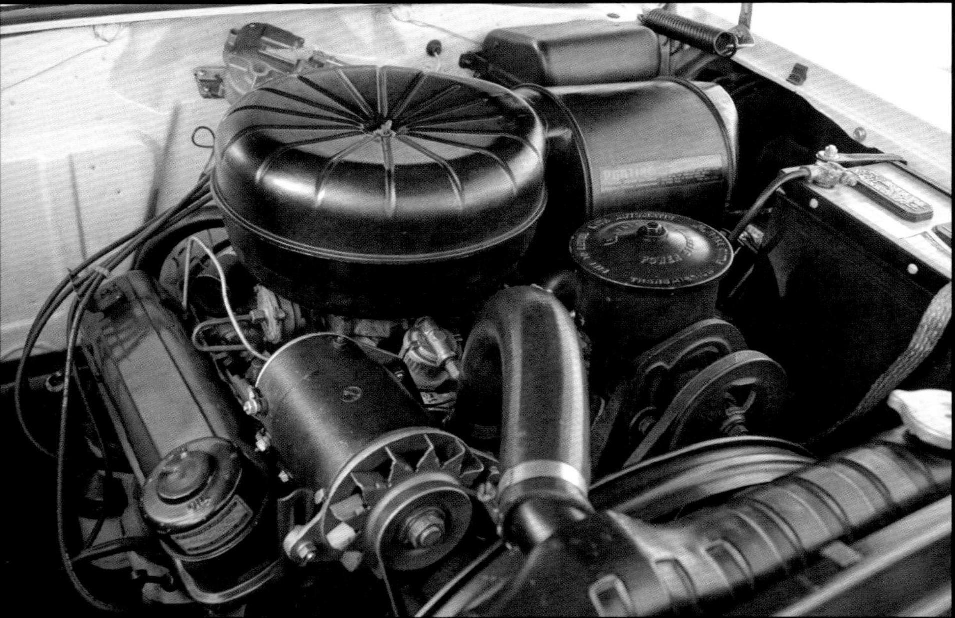

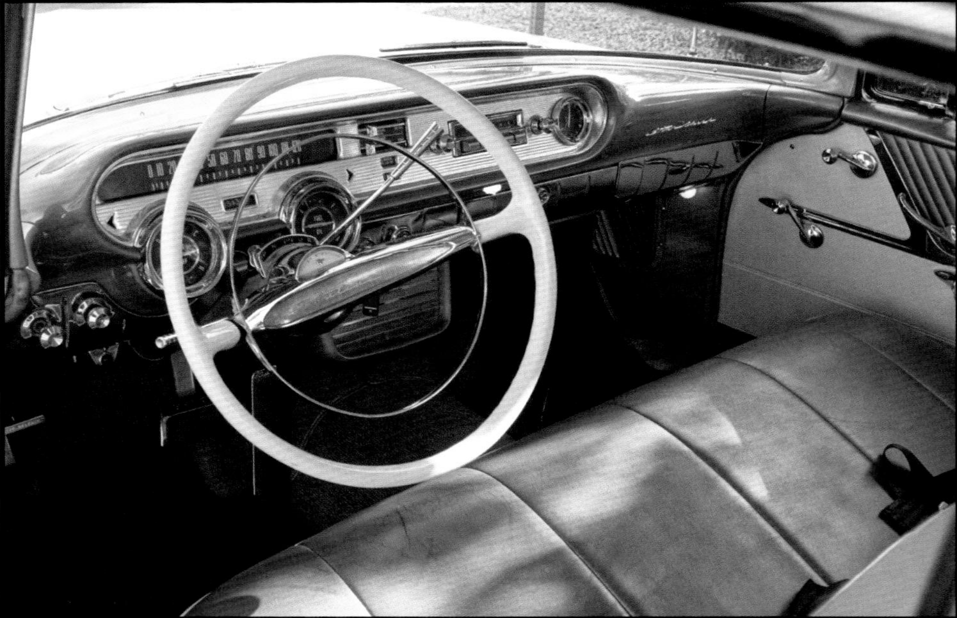

Consul Convertible Coupe

Ford Motor Company had to roll up its sleeves after World War II. Just as its domestic cars needed a thorough redesign for the postwar market, so did cars from Ford's English branch.

Like their American cousins, British Fords still had transverse-leaf-spring suspension and sidevalve engines. The 1951 English Ford Consul changed that. It was perhaps an even bigger leap forward for the firm than the American '49 models. The new Consul was powered by Ford's first ohv engine, and it put the company ahead of most manufacturers by adopting unit-body construction. Perhaps the most revolutionary aspect of the Consul was the first use of MacPherson-strut front suspension in a volume car. (Ford first used MacPherson struts on its low-production French Vedette in '48.) This independent suspension combines tubular shock absorbers within coil springs in a strut tower. Today it is probably the most popular type of suspension.

Engineering was English, but styling was heavily influence by Dearborn. The Consul sold well, and a Mark II version appeared for the 1957 model year. The new car was larger and could seat six passengers. Although considered a medium-size family car in England, the 104.5-inch-wheelbase Consul was definitely compact by American standards.

In 1953, the Consul sedan was joined by a convertible. The convertible carried over to the Mark II range. Rather than build the low-volume open car itself, Ford sent sedan body shells to Carbodies of Coventry (famous builders of the London Taxi) for conversion. A distinctive feature of the convertible was a three-position top. Besides the usual full up and down positions, the top could be erected over only the rear seat, leaving the front compartment open—which British literature referred to as the "smart DeVille position."

The Mark II's four-cylinder engine grew to 104 cid and 59 bhp. The convertible weighed a tidy 2472 pounds and cost $2351 in the U.S. but, at only about $300 less than a big Fairlane 500 Sunliner convertible, sales were low.

This car is one of maybe 10 '58 Consul convertibles that Ford sold in the U.S. The three-speed stickshift of this Consul has been replaced by an automatic from a six-cylinder English Ford Zephyr.

1958 Ford Consul Convertible Coupe

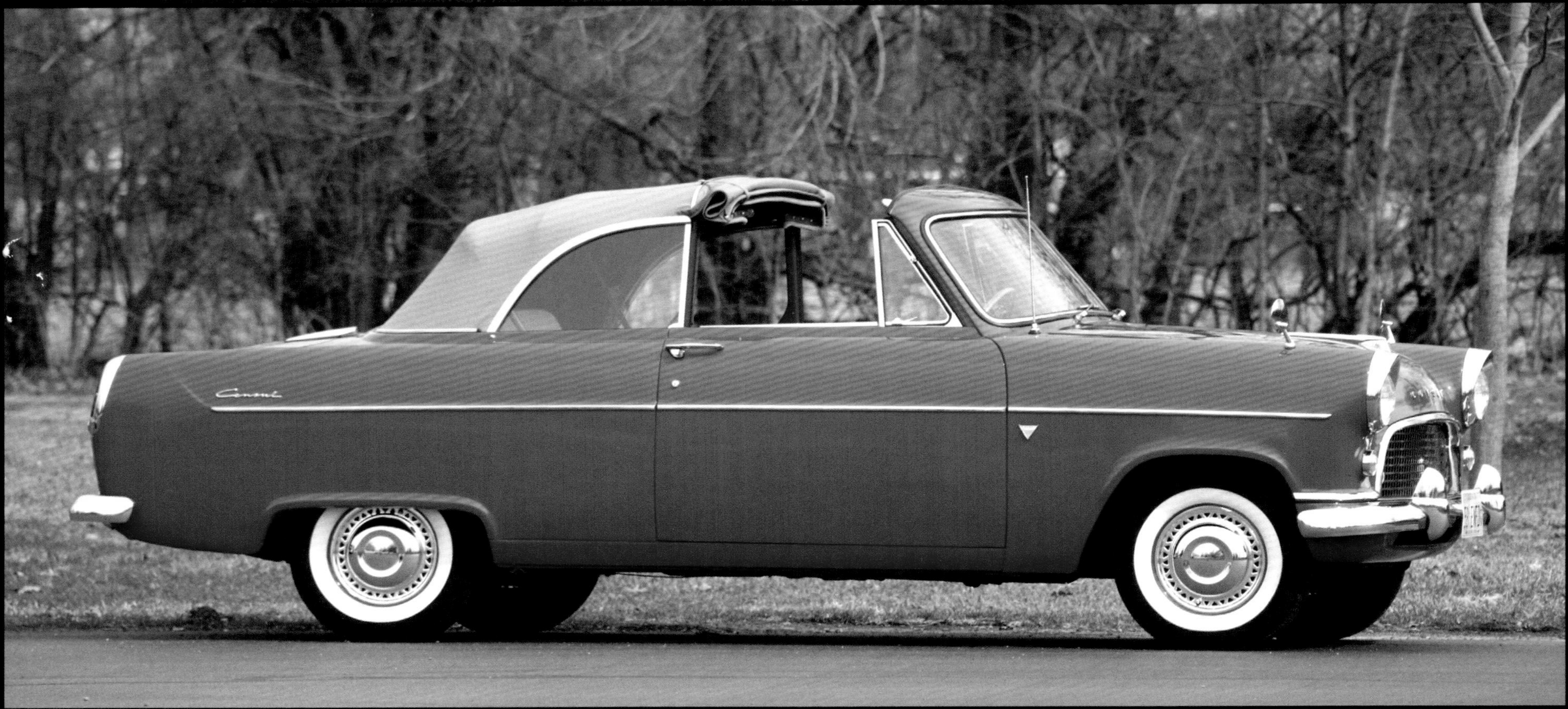

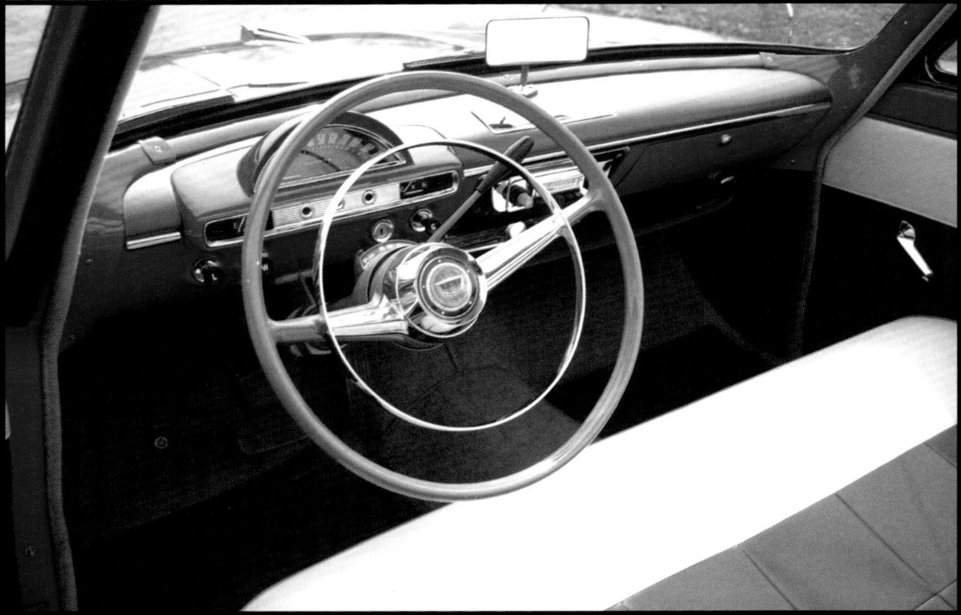

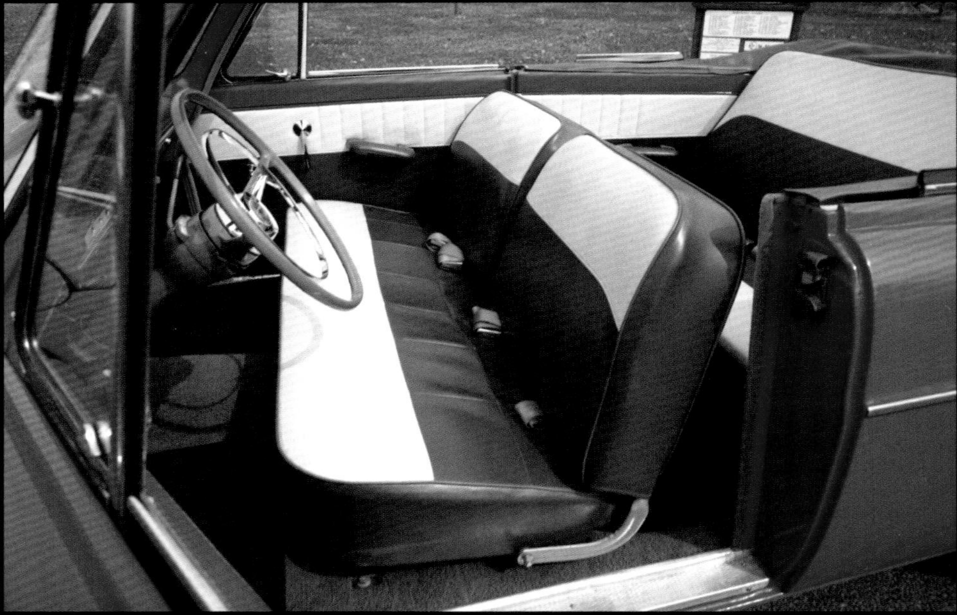

The American station wagon has had many transformations over the years. It went from being seen as a commercial vehicle to being accepted as a passenger-car style. Wood body construction gave way to steel. And though four-door convenience made the most sense for such a vehicle, two-door wagons enjoyed a brief heyday that peaked in the mid-Fifties.

Perhaps the most famous two-door wagon was the 1955-57 Chevrolet Nomad. It boasted show car-inspired roof styling and top-end Bel Air-series appointments. In the nature of their market competition of the day, anything that sales leader Chevy had, Ford wanted too. While Ford wasn't willing to create special bodywork for a custom two-door, it did fashion a premium model out of its basic Ranch Wagon.

When Ford went to all-steel wagon bodies for 1952, the Ranch Wagon was the entry-level model, albeit a popular one. Two years later, the midlevel Customline series added a Ranch Wagon companion to its four-door Country Sedan. Then, in '56, with the Nomad on the market, Ford fired back with the Parklane. It featured high-end Fairlane trim inside and out on a Ranch Wagon body. Its base price was $180 less than the Nomad's, and it outsold the fancy Chevy by about 2-to-1, though that came to just 15,186 cars.

Fords got all-new styling on a brand-new chassis for 1957. The Parklane didn't return for '57, but there was something called the Del Rio Ranch Wagon. Featuring the gold-anodized side trim of two-toned Custom 300 sedans, plus interior materials to match, it was a sort of cross between the Parklane and the former Custom Ranch Wagon.

The Del Rio carried over into 1958, but by then there was little need for it. Chevy had abandoned the specially bodied Nomad. From a healthy showing of 46,105 Del Rio Ranch Wagons for 1957, demand slumped to 12,687 of the '58s. In 1959, the Del Rio was replaced by a two-door Country Sedan, but having sold just 8663 copies, it was dropped. The last base two-door Ranch Wagon was the '61.

Ford had a new family of V-8 engines for '58, topped by a 300-bhp 352-cid mill with a four-barrel carburetor. This Del Rio has a 352, although it's a two- barrel version from later years. The transmission is a three-speed Cruise-O-Matic automatic, another new option for 1958 Fords.

1958 Oldsmobile
Dynamic 88 Hardtop Coupe

It might have been hard at the time for the loyal Oldsmobile customer to believe, but somewhere deep inside, the 1958 models were based on the same shells that had served the '57s. That's because the '58s had new looks from front to back and roof to road.

Styled under the direction of Art Ross, who had headed the Oldsmobile Studio since 1947, the 1958 Oldses broke with recent marque styling trends. The elliptical "jet-intake" grille of prior years was replaced by a full-width horizontal-bar grille and massive bumper. "Hockey-stick" side trim gave way to a projectile shape in front and horizontal strakes in the rear, which gave the unfortunate impression of being on a collision course with each other. Revised roofs featured an undivided backlight, which alleviated criticism of the three-piece window used on the '57s. About the only visual tip that this was an Oldsmobile was the continued use of high-mounted, round taillights. Wheelbase was up by a half inch (to 122.5), and '58s were 2.2 inches wider than the '57s.

The anchor of the 1958 Olds lineup was the Dynamic 88 series. (Like quad headlights and optional air suspension, the Dynamic sobriquet was new.) Powered by a standard 265-bhp "economy" version of the Olds 371-cid V-8, it came in seven models, including a $2834 Holiday two-door hardtop that drew 50,7897 orders.

This original Alaskan White-over-Banff Blue Dynamic 88 hardtop coupe is equipped with optional power steering and brakes, a radio, whitewall tires, and Hydra-Matic automatic transmission in place of the standard three-speed synchromesh transmission.

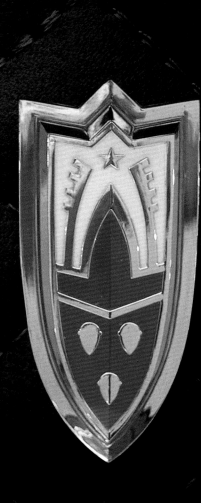

1958 Simca

Vedette Présidence Four-Door Sedan

Simca—acronym for Société Industrielle de Mécanique et Carrosserie Automobile (roughly translated, "industrial company of automobile mechanics and body")—was founded in 1935 in France. At first, Simca produced Fiat designs in the former Donnet factory at Nanterre, near Paris. When production resumed after World War II, the company steadily moved toward making cars that differed more from Fiat's offerings.

During the Fifties, Simca expanded by acquiring other companies. These included Talbot and truck builder Unic. For our story, though, the most important of Simca's acquisitions was that of Ford France. Ford's history in France dates to 1916 when a Model T assembly plant was set up at Bordeaux. After World War II, as in America, Ford France resumed production with a car based on a prewar model. The French car used a 2.2-liter V-8 engine that was very similar to the L-head "V-8/60" produced in the U.S. from 1937 to 1940.

Meanwhile, in Dearborn, the parent company had prepared a so-called "Light Ford" for the postwar market designed (by Bob Gregorie) around the V-8/60. As envisioned, the small car was to be joined by a "full-size" Ford for 1948. However, by late 1946, plans changed: The full-size Ford became the 1949 Mercury and the smaller car was transferred to Ford France, where it entered production in late 1948 as the Vedette.

The Vedette was redesigned for 1955, shifting to unit construction. The front suspension was handled by MacPherson struts and the small "flathead" V-8 was still found under the hood. This car saw short life as a Ford, because in late 1954 Ford sold its French operations to Simca. In return, Ford received shares equaling a 15.2-percent ownership stake in Simca. Vedette production continued under Simca. A 1958 redesign introduced contemporary American styling themes including a wraparound windshield and tailfins. The '58s were introduced in three states of trim: base Ariane 8, midlevel Beaulieu, and top-line Chambord.

In August 1958, Chrysler announced it was purchasing Ford's stake in Simca and taking over U.S. distribution. Then, at the 45th Paris Motor Show that October, Simca introduced a new top-of-the-line Vedette, the Présidence. The car at the show was outfitted as a mobile office suitable for top government officials. Space-age electronic extras included a radio telephone and a television set built into the back of the front seat. The Paris car had a limousine-style divider between the front and rear compartments. The production Présidence was also available with individual reclining front "Pullman" seats. Other features included black exterior paint and an externally mounted spare tire.

Like other Vedettes, the Présidence rode a 106-inch wheelbase. But at 195.7 inches long, it was 8.7 inches longer. The 2351cc "Aquilon" V-8 engine was rated at 84 bhp. It mated to a three-speed manual transmission. Overdrive and Simca's Rushmatic semiautomatic transmission were Présidence options. After some legal wrangling with existing Simca distributors, Chrysler did take over sales in the U.S. Models sold in the States included the small Aronde, the Vedette-based four-cylinder Ariane, and the Vedette—though not the Présidence. With import-car demand growing in late-Fifties America, Simca sales more than doubled to about 35,000 for '59, good for sixth place among imports.

Vedette production continued in France until 1961. Flathead-powered Vedettes, including the Présidence, were also built in Brazil until 1967. By 1970, Simca became part of Chrysler Europe. Exports to the United States continued into 1971. Chrysler sold its European operations to Peugeot in 1978, two years before the Simca name was last used.

6681 SJ 49

Corvette Convertible

Nineteen-sixty was the Chevrolet Corvette's breakthrough year. It was the year when the fiberglass-bodied sports car cracked the 10,000-unit barrier, the magic number that meant the 'Vette was finally earning its keep at General Motors.

When the Corvette was introduced late in the 1953 model year, production forecasts of at least 10,000 units for the following season were made. The projections proved wildly optimistic, however, and by '55, the Corvette was nearly extinct. Then faster, prettier, and more comfortable 'Vettes started coming off the St. Louis assembly line beginning in 1956. Demand rose in response to the improvements to the point that 10,261 of the '60 models were assembled, an increase of about 600 cars compared to 1959.

There was more to the 1960 breakthrough than just healthier sales. Sure, there seemed to be a few more Corvettes on Main Street, but they were turning up lots of other places, too. Briggs Cunningham took three 'Vettes to LeMans, one of which finished eighth in the GT class. That October, television viewers got their first look at *Route 66*, a weekly series in which Martin Milner and George Maharis portrayed Tod Stiles and Buzz Murdock, two young adventurers seeing the USA in their two-seat Chevrolet.

Externally, the 1960 'Vette was virtually identical to the '59, itself a slightly cleaned-up version of the 1958 model. Inside, vertical seat pleats replaced the '59's horizontal pleats.

Underneath, though, the 1960 gained a rear sway bar, got a heftier front anti-roll bar, and received an added inch of rear suspension rebound. These changes added up to improved ride and handling.

Outputs of the 283-cid V-8s used in Corvettes were carried over unchanged for 1960, but the 250- and 290-bhp fuel-injected versions gained a freer-flowing plenum. (Rare—but troublesome—aluminum heads were newly available for "fuelie" Corvettes.) Manual gearboxes were required with all injected engines.

Eight paint colors were offered for 1960, including a new metallic Cascade Green. Equipped with a 290-bhp engine and optional four-speed transmission, this restored Corvette is one of just 65 painted in the green-and-white scheme.

1960 Chevrolet Corvette Convertible

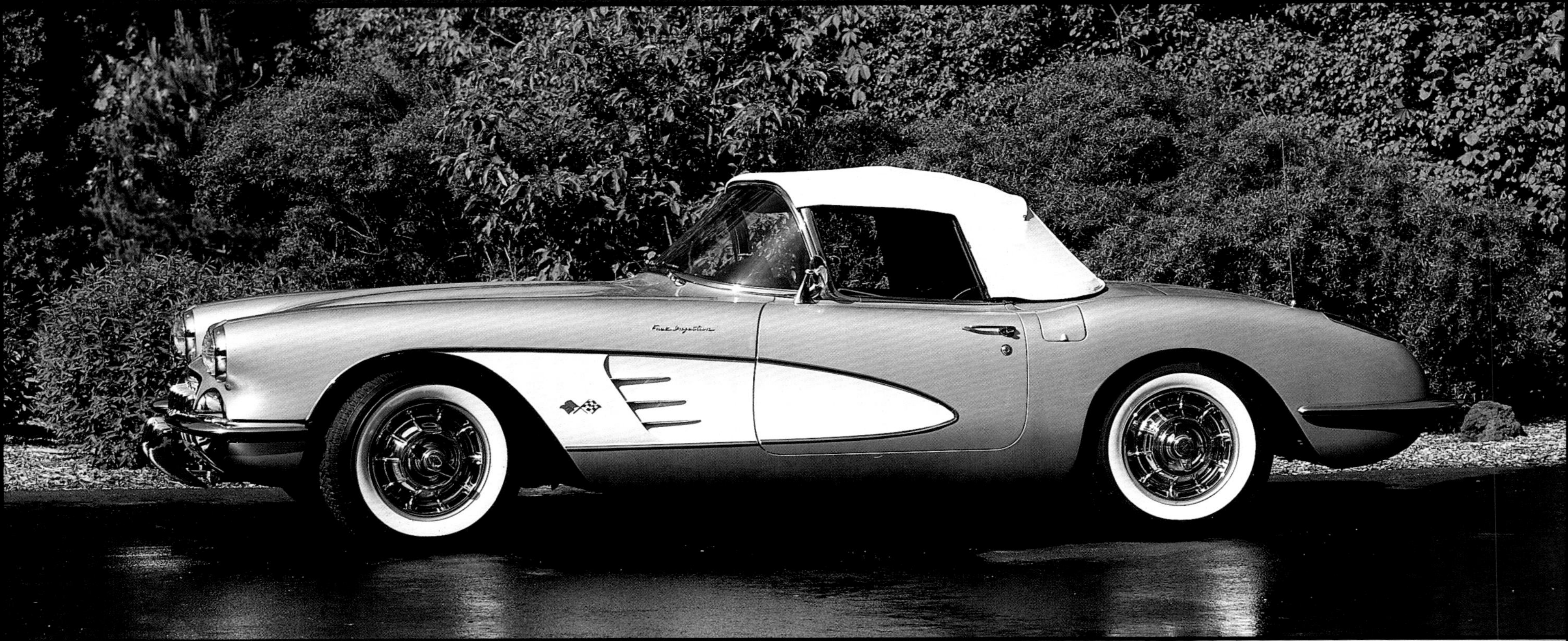

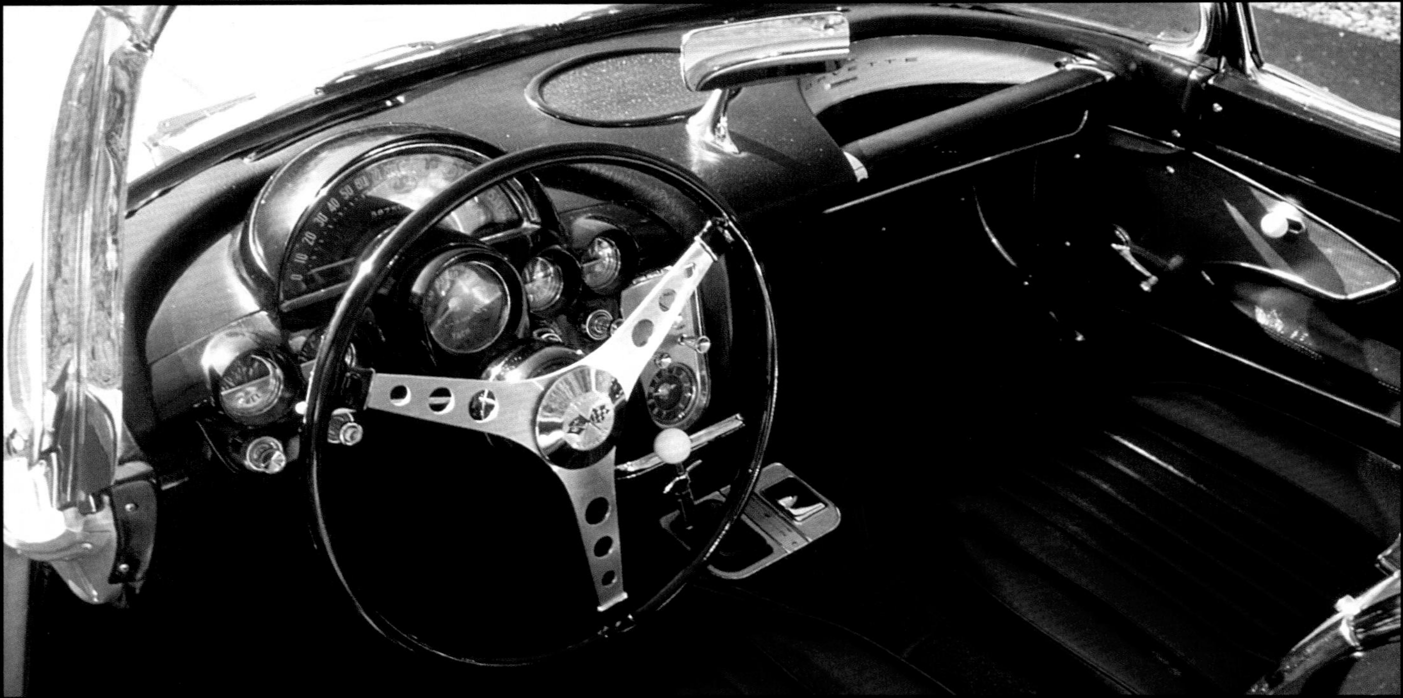

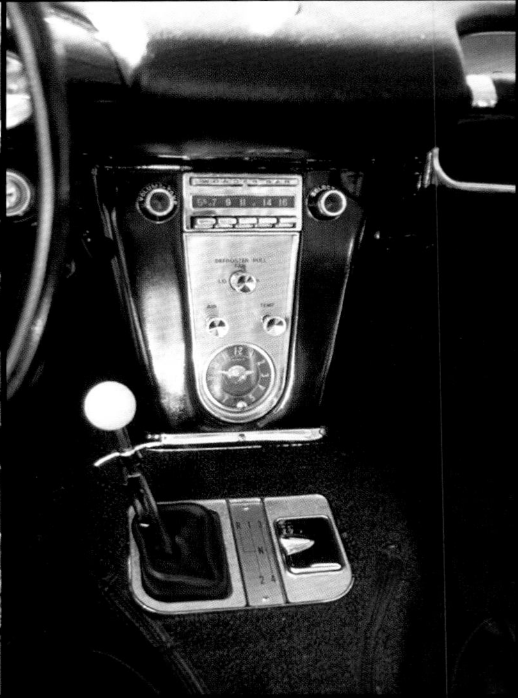

1961 Plymouth
Belvedere Four-Door Sedan

Plymouth spent the early Sixties in crisis mode. Sales of the 1960 Plymouths had been disappointing. Although the 1961 car was a continuation of the 1960 body shell, it got a complete makeover. Only the roof and doors were carried over—all other sheetmetal was new.

The tailfin fad had run its course and Chrysler styling chief Virgil Exner pruned them away for 1961. That doesn't mean that Exner had suddenly become conservative. Taillights were housed in pods and the front-end styling was controversial, to say the least.

Inside, the dashboard was restyled. The band-type speedometer remained in a pod mounted on top of the flat dash, but was no longer gear driven. Instead a unique magnetic system operated the speedometer, and was said to be more accurate. An optional clock was mounted under the speedometer and was flanked by temperature and fuel gauges. A blank face greeted customers who didn't pay extra for the timepiece. The extra-cost heater had push button controls.

Ultimately, sales slid even further. It's unfortunate that Plymouth styling didn't appeal to more buyers. Under the skin, Plymouths were good cars. Unibody construction offered better rigidity than the previous body-on-frame Plymouth. The torsion-bar front suspension gave better handling than competitors Ford and Chevrolet, but was still smooth riding. V-8 horsepower ranged from 230 to 375 in 1961. Then, too, for economy-minded buyers, there was a highly regarded six-cylinder engine.

Chrysler Corporation introduced its compact Valiant in 1960, powered by a new ohv six that replaced a flathead six with roots that went back to the Thirties. To fit under the Valiant's low hood, the inline six was inclined 30 degrees to the right and became commonly known as the "Slant Six." Plymouth often labeled it "30-D Economy Six."

Tilting the engine did more than help it clear the hood. It also created room for long intake-manifold runners that resulted in more efficient breathing. Valiants used a 170-cid version of the Slant Six, while full-sized Plymouths, which also adopted the engine, had a 225-cid unit with 145 bhp.

Slant Six performance and fuel economy were both good for its size, and over time the engine gained a reputation for bulletproof durability. The '61 Plymouth brochure noted that the full-size Plymouth six "walked off with top honors for its class in the 1960 Mobilgas Economy Run." Besides being thrifty, the Plymouth six also produced 10 more horsepower than the similarly sized sixes from Ford and Chevrolet.

This car is a midline Belvedere sedan powered by a Slant Six mated to a three-speed manual transmission. It has optional power brakes, but not power steering. Base price was $2439 and 40,090 Belvedere sedans—six and V-8—were manufactured. Only the base Savoy four-door sedan had a bigger run.

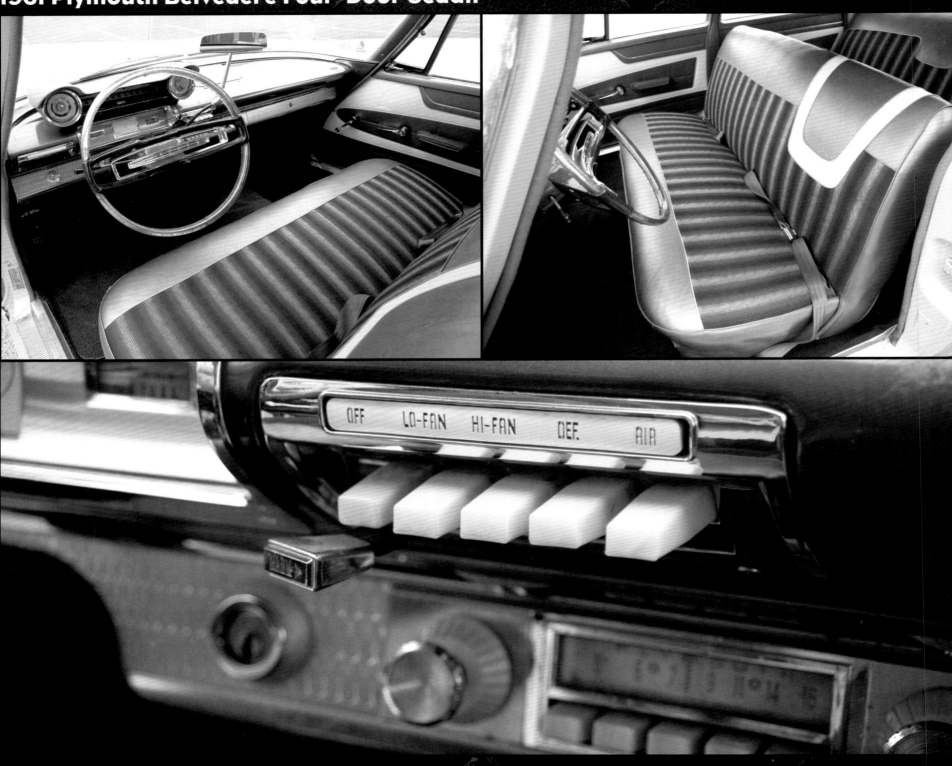

OFF   LO-FAN   HI-FAN   DEF.   AIR

1962 Chevrolet
Corvair Monza Station Wagon

Things would never be better for the Chevrolet Corvair than they were in 1962. True, more power, critical chassis upgrades, and an attractive restyle were all in the offing for General Motors' controversial compact. But '62 brought exciting new models and enhanced performance, and production of 292,531 cars would stand as the high point in the ten-year run of Chevy's answer to the Volkswagen Beetle.

A convertible and a turbocharged Spyder option were added to the Corvair's sporty and popular Monza series for '62. So, too, was a station wagon. The wagon made its debut in 1961 as part of the midlevel 700 series. The six-passenger Lakewood, as it was called, offered 58 cubic feet of cargo space on a load floor that was nearly 6.5 feet long with the second-row seats folded down. A handy one-piece liftgate lent easy access to the cargo hold, but the floor sat well above bumper height to make room for the Corvair's aluminum-block, horizontally opposed, air-cooled six-cylinder engine.

The Monza wagon started at $2569 and was trimmed like the other models in the series (though bucket seats, which gave other Monzas a sporty identity, were an option). The 145-cid engine made 80 bhp with the standard three-speed manual transmission; 84 bhp with the extra-cost Powerglide automatic. Powertrain options included a four-speed stick and a "Super Turbo-Air" engine with output raised to 102 bhp.

Monza glamour didn't seem to rub off on the station wagon. Just 2362 were made in '62 (plus another 3716 in 700 trim), and the body style was dropped.

Among the rare survivors of that handful of Monza wagons is this Twilight Blue example seen here. Power is provided by the 102-horse Super Turbo-Air engine, which is mated to the Powerglide. Other extra-cost items on this car include its front bucket seats, tissue dispenser, wire wheel covers, roof rack, bumper guards, and side-window "ventshades."

1965 Ford
Falcon Squire Station Wagon

The Ford Falcon was Robert McNamara's baby. A practical "numbers guy," McNamara hated waste and excess. The Edsel went against his core beliefs with its large size, superfluous decoration, and the fact that it competed with existing Ford and Mercury products. As the Edsel was failing, McNamara was campaigning for a compact Ford.

By then McNamara was vice president of Ford's North American vehicle operations. He wanted something like the imports and the Rambler American that proved popular during the 1958 recession. He was not alone in his desire for an economy compact. General Motors and Chrysler Corporation were also planning similar cars.

The Big Three introduced their new small cars for the 1960 model year. The Chevrolet Corvair was the most innovative and Chrysler's Valiant had the most powerful engine and expressive styling, but Ford's Falcon had the lowest starting price. Unlike Corvair, Falcon was simply engineered; unlike Valiant, it had conventional styling. The small Ford was what the public wanted. Falcon's first-year production total of 435,676 was more than Corvair and Valiant combined.

Ford engineers produced a car that was light and had good interior room for its size. McNamara made sure it was economical to build and sell. The first Falcons were sparsely equipped and offered few options—some thought the cars harked back to Ford's Model T.

After Falcon's successful launch, McNamara became president of Ford Motor Company, only to leave a few months later to be Secretary of Defense in the Kennedy Administration. With his departure, Falcon came under Lee Iacocca's influence. Iacocca came up from sales and knew what buyers wanted. McNamara created a car that was economical to build and own. Iacocca made it more appealing. Better-trimmed models were added and the option list expanded—both of which increased profits.

The economical but underpowered base six-cylinder engine was joined by a more powerful six and, later, a V-8. A sporty Sprint model was added, a foreshadowing of the Falcon-based Mustang, a runaway success that wouldn't have been possible without McNamara's no-nonsense Falcon.

This 1965 Ford Falcon Squire station wagon was built during the final year of the Falcon's first generation. By then Falcon had a more upmarket appearance and features. The $2665 Squire was the best trimmed Falcon wagon with its simulated-wood body adornments.

This example has the extra-cost 200-bhp 289-cid V-8 and Cruise-O-Matic three-speed automatic transmission. Wagons weighed about 300 pounds more than sedans, so the V-8 aided both performance and hauling capabilities. The car also has optional power steering and air conditioning—the kinds of things Iacocca felt shoppers wanted. Although a compact 190 inches long, the Falcon wagon had a generous 77.9-cubic-foot cargo volume.